Writing Science in Plain English

Writing
Science
in Plain
English

ANNE E. GREENE

THE UNIVERSITY OF CHICAGO PRESS

Chicago and London

The University of Chicago Press, Chicago 60637

The University of Chicago Press, Ltd., London

© 2013 by Anne E. Greene

All rights reserved. Published 2013.

Printed in the United States of America

27 26 25 24   14

ISBN-13: 978-0-226-02637-4 (paper)
ISBN-13: 978-0-226-02640-4 (e-book)

Library of Congress Cataloging-in-Publication Data

Greene, Anne E.
Writing science in plain English / Anne E. Greene.
pages. cm. — (Chicago guides to writing, editing, and publishing)
Includes index.
ISBN 978-0-226-02637-4 (pbk. : alk. paper) — ISBN 978-0-226-02640-4 (e-book)
1. Technical writing. 2. Communication in science. 3. English language—Style.
I. Title. II. Series: Chicago guides to writing, editing, and publishing.
T11.G6814 2013
808.06′65—dc23

2012043145

♾ This paper meets the requirements of ANSI/NISO Z39.48-1992 (Permanence of Paper).

To the memory of my mother, M. C. LINTON,
who wanted me to write.

True Ease in Writing comes from Art, not Chance

As those move easiest who have learn'd to dance.

ALEXANDER POPE

Contents

Acknowledgments

I feel honored to recognize all the support I've received while writing this book. I have never worked with so many interesting, intelligent people.

I am deeply indebted to the people whose writing I used anonymously for the examples and exercises. I felt strongly that only by using real examples could I make a convincing case for better scientific writing. I was privileged to have several colleagues and friends read the first draft and encourage me to keep going: Leo Lott, Ben Roberts, and particularly Richard Wydick. You helped give the fledgling wings. I also want to thank faculty and instructors at several universities and colleges who discussed scientific writing and convinced me the book was needed: John Alcock, Chris Beachy, Gregory Beaulieu, Harold Bergman, Steve Buskirk, Michael Canfield, Kerry Foresman, Scott Freeman, Jon Herron, Bruce Lyon, Mike Minnick, Kevin Murray, Ron Royer, Anna Sala, Brian Sullivan, Bret Tobalske, Scott Wetzel, and David Wilcove. Here at the University of Montana, the support of my colleagues was remarkable. Many people helped me locate good writing, answered questions, checked revisions, and read drafts. Sincere thanks to Barry Brown, Jay Bruns, Kelly Dixon, Doug Emlen, Aaron Flesch, Kathy Garramone, Mark Grimes, Jesse Hay, Dick Hutto, Charles Janson and the staff of the Division of Biological Sciences, Steve Lodmell, John Maron, John McCutcheon, Johnnie Moore, Dan Pletscher and my students and colleagues in the Wildlife Biology Program, Frank Rosenzweig, Holly Thompson, Kelly Webster, and Scott Wetzel. I am grateful to Betty Sue Flowers of the University of Texas for kindly offering to share her ideas.

Writing a book about writing can be fraught with problems when the author doesn't follow her own advice. Art Woods has my undying gratitude for keeping me honest as he edited each chapter and made invaluable comments. Ben Roberts deserves a special note of thanks for directing me and my manuscript to the University of Chicago Press, where the editorial staff skillfully guided me through the publishing maze. I am deeply grateful to Christie Henry and Paul Schellinger for responding positively to my idea, to Dawn Hall for her excellent suggestions, and particularly to Mary Laur, who settled in for the long haul and did a superb job as my editor and trusted colleague.

My family was my best resource through it all—my two daughters, Allison and Robin, supported me in countless ways—but my husband, Erick, deserves my greatest thanks. He sustained me from the outset, read my drafts, gave me comments, counseled my decisions, fixed my computer, cooked my meals, and took me flying. I can't thank him enough.

CHAPTER 1

Why Write Science in Plain English?

"Do you dread reading the scientific literature?" I ask on the first day of my scientific writing class. My students, mostly honors undergraduates majoring in biology, roll their eyes and nod their heads. They all agree—reading science papers is hard. When I ask why, the responses are telling: "Reading papers puts me to sleep," or, "I have to read them three or four times before they make sense," or, "They make me feel stupid." Why should intelligent, motivated students have difficulty reading the scientific literature? The answer is because most of it is poorly written.

If you are an undergraduate student, perhaps you have the same problem with your assigned readings. If you are a graduate student, a postdoctoral fellow, or an established scientist, perhaps you have heard similar complaints from your students, or had these thoughts yourself.

The truth is, many journal editors and senior scientists believe that unclear scientific writing is a serious problem. Peter Woodford, former president of the Council of Science Editors, described the poor writing he saw in journals as "appalling."[1] Leslie Sage, senior editor of physical sciences at *Nature*, wrote, "It is a sad commentary . . . that many of the 'crank' papers submitted to *Nature* are actually better written—from a purely stylistic point of view—than many of the professional papers."[2] Harold Heatwole, editor of *Integrative and Comparative Biology*, concluded, "The standard of writing in current scientific journals has reached an all-time low, in terms of both poor grammar and imprecise communication."[3] Many senior scientists who have written on the subject agree with David Porush that scientific writing is "unnecessarily dry, difficult to read, obscure, and ambiguous."[4] They urge scientists to write more clearly, with more directness and precision,[5-8] in a style Anthony Wilson calls "plain, simple English."[9]

Yet scientific writing, while exploding in quantity, is not improving in quality.[10] In a survey of 22 journals on atmospheric science, a measure of clarity of the articles was either holding steady or declining.[11] As recently as December 2011, the chief scientific editor of *Science Signaling* described the computational results in many manuscripts as "obscure, convoluted, jargonistic, or impenetrable."[12]

Why does this epidemic of poor scientific writing matter? One reason is it hinders the flow of ideas across disciplines. As science becomes more specialized and the writing more complex, specialists in different fields struggle to understand one another.[13] Poor writing also makes it more difficult to apply discoveries from one field to another, a cross-fertilization that has advanced scientific discovery in the past.[11,14] One scientist recently suggested that unclear writing hinders the scientific process itself.[12]

In addition, poor scientific writing is partly to blame for the decline in science literacy in the United States and the long-standing communication gap between scientists and the general public.[4,9,15–17] If we are to solve the profound problems facing our nation and the world, decisions must be shaped by science-literate citizens and lawmakers.[18] But in a recent poll conducted by the Pew Research Center for the People and the Press, 85% of scientists surveyed say public ignorance of science is a major problem. About half the Americans surveyed disagreed that human activities are causing global climate change, and almost a third don't believe in human evolution.[19] To help close the rift, the president of the American Association for the Advancement of Science (AAAS), Peter Agre, urged "every scientist and engineer to make their work both beneficial and understandable" to the general public.[18]

Younger scientists may be our best hope. In their book *Unscientific America*, Chris Mooney and Sheril Kirshenbaum describe a crisis in communication between scientists and "everyone else" that could be improved by training "Renaissance scientists" who can communicate more effectively.[20] Similarly, the CEO of the AAAS and executive publisher of the journal *Science*, Alan Leshner, believes that young scientists should be trained in "public communication," and that scientists who share their research with a broad audience should be rewarded.[21]

But poor writing sets a bad example for young scientists. If you are a newcomer to the field, you probably imitate the writing you read in professional journals, a common enough practice in any profession, but one that guarantees that poor writing persists.[4,8,22] Even if you are an established scientist, your writing style was probably influenced by your major professors or advisors, few of whom were trained to write clearly or to instruct others to do so.[7,17] One consequence of this is the feedback most science students receive on their writing varies enormously.

The good news is you can write science in plain English by applying a relatively short list of principles developed for professional writers by Joseph Williams in his book *Style: Toward Clarity and Grace*.[22] These principles are based on linguistic theory about what readers look for when they read complex, unfamiliar information. The list is surprisingly simple: readers look for a story about characters and actions; for strong verbs close to their subjects; for old information at the beginnings of sentences and new information at the ends; and for specific kinds of information in predictable places in paragraphs and documents.

Williams's principles and their linguistic history are at the heart of this book. Most other books on scientific writing focus on *what* scientists write; they describe how to prepare a thesis, a grant proposal, a research paper, and a review article; many include instructions on data presentation, formatting, and citation styles; some cover how to give an oral presentation and how to prepare a poster.[5,8,23,24] They don't concentrate on *why* scientific writing is so hard to understand or *how* to improve it.

This book dispenses with information about what scientists write and focuses entirely on how to write clearly and comprehensibly. The principles it describes will help improve everything you write, whether it is a lab report, a grant proposal, a research paper, or a press release. At what stage in the writing process you use the principles is up to you. You might use them to revise a first draft, or once you are familiar with them, you might incorporate the principles as you write. Just remember that at some stage you must adjust your writing so that it gives your readers what they need to understand you.

Before you begin to write, you must choose your audience, register, and tone. These topics are discussed in Chapter 2. The remaining chapters describe the principles, using good and bad examples of real scientific writing. Once you understand each principle, you can practice it by doing the exercises at the end of the chapter. Then compare your results with those in the Exercise Key in Appendix 2.

Throughout the book, I use some common grammatical terms that refer to parts of speech and basic sentence structure. If you are unsure of these terms or need a quick refresher on grammar, refer to Appendix 1. It's important that you understand the terms because I use them to explain how the principles work, and they will help you apply the principles to your own writing.

Because many problems with scientific writing are common to all disciplines and at all levels, these principles will help whether you are a geologist, chemist, physicist, biologist, or social scientist, and whether you are a first- or fourth-year undergraduate, a graduate student, a postdoctoral fellow, or a professor.

Certainly, the merit of your scientific writing rests as much on content as on style. Equally important are the questions, hypotheses, experimental designs, and interpretations you describe. However, if you cannot clearly communicate these things to your readers, what is the point?

REFERENCES
1. Woodford, F. P. Sounder thinking through clearer writing. *Science* **156**, 743–745 (1967).
2. Sage, L. in *Astronomy Communication* (eds Heck, A. & Madsen, C.) 221–225 (Kluwer Academic Publishers, 2003).
3. Heatwole, H. A plea for scholarly writing. *Integr. Comp. Biol.* **48**, 159–163 (2008).
4. Porush, D. *A Short Guide to Writing about Science* (Longman, 1995).
5. Ebel, H. F., Bliefert, C. & Russey, W. E. *The Art of Scientific Writing: From Student Reports to Professional Publications in Chemistry and Related Fields* (Wiley-VCH, 1987).
6. O'Connor, M. *Writing Successfully in Science* (HarperCollins, 1991).
7. Alley, M. *The Craft of Scientific Writing* 3rd edn (Springer, 1996).
8. Schultz, D. M. *Eloquent Science: A Practical Guide to Becoming a Better Writer, Speaker, and Atmospheric Scientist* (The American Meteorological Society, 2009).

9. Wilson, A. *Handbook of Science Communication* (Institute of Physics Publishing, 1998).

10. Wells, W. A. Me write pretty one day: How to write a good scientific paper. *J. Cell Biol.* **165**, 757–758 (2004).

11. Geerts, B. Trends in atmospheric science journals: A reader's perspective. *Bull. Am. Meteorol. Soc.* **80**, 639–651 (1999).

12. Yaffe, M. B. The complex art of telling it simply. *Sci. Signal.* **4**, doi: 10.1126/scisignal.2002710 (2011).

13. Gould, S. J. Take another look. *Science* **286**, 899 (1999).

14. Sand-Jensen, K. How to write consistently boring scientific literature. *Oikos* **116**, 723–727 (2007).

15. White, F. D. *Communicating Technology: Dynamic Processes and Models for Writers* (HarperCollins, 1996).

16. Sabloff, J. A. Distinguished lecture in archeology: Communication and the future of American archaeology. *Am. Anthropol.* **100**, 869–875 (1999).

17. Barrass, R. *Scientists Must Write* 2nd edn (Routledge, 2002).

18. Lempinen, E. W. (ed) Science leaders urge new effort to strengthen bonds with public. *Science* **327**, 1591 (2010).

19. Pew Research Center for the People and the Press. *Scientific Achievements Less Prominent Than a Decade Ago: Public Praises Science; Scientists Fault Public, Media.* Available at http://www.people-press.org/reports/pdf/528.pdf (2009).

20. Mooney, C. & Kirshenbaum, S. *Unscientific America: How Scientific Illiteracy Threatens Our Future* (Basic Books, 2009).

21. Leshner, A. I. Outreach training needed. *Science* **315**, 161 (2007).

22. Williams, J. M. *Style: Toward Clarity and Grace* 5th edn (Univ. Chicago Press, 1995).

23. Hofmann, A. H. *Scientific Writing and Communication: Papers, Proposals, and Presentations* (Oxford Univ. Press, 2010).

24. Pechenik, J. A. *A Short Guide to Writing about Biology* 7th edn (Longman, 2010).

CHAPTER 2

Before You Write

It's usually a good idea to plan ahead, and with writing, planning ahead can make the difference between success and failure. Before you write, decide who you are writing for, how formal you should be, and the attitude you want to project. These decisions will help determine if your writing is clear and interesting. Often, scientists reflexively favor a dry, abstract, and unvarying style, but we can do better by considering our audience, register, and tone before we write.

Audience

The most important first step is to envision your audience. Who will read your report, paper, thesis, or textbook? Your audiences could include family or friends, interested nonscientists, or other scientists who may or may not share your discipline. Most likely, some of your readers will know less about the subject than you do, so put yourself in their shoes. They are trying to understand you, but *don't know what you know*. Help them by making your writing as clear as possible. If you are unsure of your audience, err on the conservative side. Write for the reader who may be least informed. By doing so, you won't confuse anyone, and you will reach more readers.

As a student, you usually have an audience of one—your professor. However, in the real world, you will be writing for many different audiences, and your success will often depend on whether you communicate clearly to each of them. So when you write a paper or a lab report, envision a larger audience—one that is not as well informed as you are—and write for them. Writing about your subject clearly and simply will also show your professor that you understand the course concepts.

Writing with your audience in mind informs the principles in the rest of the book. It also determines two other qualities of your writing: your register and your tone.

Register

Register describes where your writing falls on the continuum from informal to formal. Remember that as your writing becomes more formal, it becomes harder to understand, particularly for readers who don't know your subject. The examples below (except one) describe a similar subject—the mating habits of porcupines—but are written in four different registers used in scientific writing: informal, popular, conventional, and abstract.[1]

1. INFORMAL REGISTER

Have you ever wondered, "How the heck do porcupines manage to mate with all those spines everywhere?" Well, the answer to that question is pretty hard to figure out because porcupines are hard to see at the best of times, but it's almost impossible when they're courting. It turns out that the whole affair is up to the woman. When she is ready to become pregnant, she produces a very strong odor that can drive the male porcupine crazy!

We often use the informal register with family and friends. It is conversational and often emotional. It assumes familiarity between writer and reader and can frame a good story. In the above example, phrases such as, *how the heck* and anthropomorphisms such as *the whole affair is up to the woman*, make this writing unsuitable for most scientific purposes. However, in some situations, small doses of informal register can transmit spirit and enthusiasm. The following excerpt is from a letter to undergraduates written by the directors for the Research Experience for Undergraduates (REU) Program at the University of Wisconsin–Milwaukee.

Thank you for your interest in our National Science Foundation Ocean Sciences Research Experience for Undergraduates (REU) Program. . . . We expect about 80 applications for 9 fellowships, but don't let that put you off. All of life is like that, and you can't get it if you don't try! If you have a decent academic record and

can write an intelligible and personalized statement of interest, you stand a good chance of success.

For most students, this kind of writing—a mixture of informal and more formal registers—tells them these scientists are professional but also know how to relate to their audience.

2. POPULAR REGISTER

Porcupines are arboreal creatures and in the Nevada region, they live and mate in thick riparian vegetation in which it is impossible for researchers to move quietly. So, although Sweitzer has come close to catching the creatures mating, he has had to settle for stumbling upon the pairs that seem to be on the verge of reproduction—animals that provide only indirect hints about how porcupines find and pick mates. But these clues have been sufficient for Sweitzer along with fellow researcher Joel Berger of the University of Nevada, Reno, to put forward a theory that has earned them some notoriety in the select circle of experts who study this creature.

This register is typical of popular science magazines that are written for a broad audience. Written pieces using this register often tell a story, in this case about trying to find breeding porcupines. Characters that readers can visualize play important roles, such as porcupines, Sweitzer, and Berger do in this example. The writing is clear and easy to understand, with few technical terms.

3. CONVENTIONAL REGISTER

I tracked the movements of North American porcupines (*Erethizon dorsatum*) in the Great Basin of northwestern Nevada. I related these movements to breeding activities during the late summer and fall of 1991 and 1992. Male porcupines are polygamous and defend several females, and I hypothesized that (1) competitively dominant males would have larger home ranges than both subordinate males and adult females, and (2) the size of home

ranges of adult males would vary and be positively correlated with breeding success.

The conventional register is characteristic of clearly written journal articles, theses, and proposals directed at a broad scientific audience. It is more formal than the previous registers, but still clear. It tells a story with identifiable characters (*I* and *porcupines*) that do things (*track, relate, defend,* and *hypothesize*). It features many verbs in active voice. It is emotionally neutral, and it assumes the reader is familiar with some technical terms (*polygamous, dominant* and *subordinate, hypothesized,* and *correlated*).

4. ABSTRACT REGISTER

The assessment of strong directional tendencies of the North American porcupine (*Erethizon dorsatum*) in the Great Basin of northwestern Nevada was made in relation to sex-specific behavioral heterogeneity during the late summer and fall periods of 1991 and 1992. A mate-defense polygynous mating system was exhibited, and it was hypothesized that (1) comparatively larger home ranges would be defended by competitively dominant males in comparison to the home ranges of subordinate males and females and (2) male home range size variation would be positively correlated with reproductive success.

Most of what scientists read and write every day is in the abstract register. It is unclear, wordy, pompous, and dull. The main character from the example of conventional register, *I*, has disappeared. The action that character did, *tracked*, has been converted to, *the assessment*, an abstract term that gives the impression of action, that apparently takes place without anyone doing it. The story element is gone. Many of the active verbs in the conventional example have been changed to passive verbs. The number of technical terms has increased, and long strings of nouns have appeared. In the previous example, the "conventional" writer describes the *movements of North American porcupines*, while the "abstract" writer describes *strong directional tendencies of the*

North American porcupine; the "conventional" writer relates movements to *breeding activities*, while the "abstract" writer assesses them *in relation to sex-specific behavioral heterogeneity*; the "conventional" writer describes how *the size of home ranges of adult males would vary*, while the "abstract" writer describes *male home size variation*. The message written in conventional register is clear; why muddy the waters with abstractions and technical terms? By doing so, writers risk confusing their readers, or putting them to sleep.

Of the two registers unsuited to most scientific writing—the informal and the abstract—the abstract is far more common, and the one you should guard against most. Abstract writing disguises your message and confuses your readers, especially those who are less informed.

Tone

Tone is the writer's attitude toward himself, his subject, and his audience. It can range from arrogant to dismissive, timid to confident, energetic to dull, cynical to optimistic. Choosing the right tone influences how the audience feels about you and your subject. In scientific writing, we try to keep our tone neutral, but since most of our writing is persuasive (we want to convince our readers to accept a particular view or hypothesis), we should adopt a tone that projects confidence rather than doubt. Nowhere is this more important than in proposal writing, where self-assurance and enthusiasm are crucial. Consider the differences between the following two passages, one of which was excerpted from a successful grant proposal to the National Science Foundation. Can you tell from the tone which one it is?

1. Horned beetles could provide an opportunity to combine studies of trait development with experiments looking at sexual selection and the evolutionary significance of enlarged male weapons (horns). After almost ten years of research, the PI may now have the opportunity, if funded, to piece together disparate parts of the research program, offering opportunities to train young scientists, and possibly providing a picture of the evolution of unusual animal shapes.

2. Horned beetles provide an unusual opportunity to combine studies of trait development with experiments exploring sexual selection and the evolutionary significance of enlarged male weapons (horns). By building on almost ten years of research directed towards this goal, the PI now has the opportunity to forge a truly integrative research program, offering unique possibilities for inspiring and training young scientists, and providing a comprehensive picture of the evolution of some of nature's most bizarre animal shapes.

The tone of the first example is weak and tentative. Verbs such as *could provide, may have,* and *possibly providing* combine to make the writer sound unsure. Expressing doubt in phrases such as *if funded* and describing the research as having *disparate parts* doesn't help.

In the second example, the writer conveys a sense of excitement and confidence. Notice how subtle changes in words and phrases change the tone. Adjectives and adverbs such as *unusual, truly integrative, unique, comprehensive,* and *most bizarre* make this writing convincing. The writer also gives the reader a sense of urgency when he states that after ten years of research he *now has the opportunity to forge a truly integrative research program.* This excerpt came from the funded proposal, and undoubtedly, tone played an important part in its success.

REFERENCE
1. Modified from Joos, M. The five clocks. *Int. J. Am. Linguist.* **28**, 9–62 (1962).

CHAPTER 3

Tell a Story

Telling stories is a unique characteristic of being human. We use stories to develop our sense of the world—its past, present, and future—as well as our place in it. No other kind of prose communicates information so efficiently. How long we have been telling stories is unknown, but traces of symbolic art that hint at storytelling may date to at least 70,000 years ago.

Many scientists see little connection between communicating their science and telling stories. They think of stories as made-up, while science is based on fact. However, to most writers, "story" simply describes a powerful way to communicate information to an audience. Recent research has shown that our brains are wired to recognize stories with a particular structure, one that features characters and their actions, and information presented this way becomes compelling and memorable.[1] Scientists can use these same elements of stories—characters and actions—to write about the real world with the same desirable results. Writing stories about science doesn't mean making it up or dumbing it down. Rather, we can hang complex ideas on the scaffolding of good, simple stories and make our science as exciting to our audience as it is to us.

Make Characters Subjects and Their Actions Verbs

How do we write stories about science? All stories feature *characters* that *act*. You can fulfill these two fundamental requirements of a story by choosing characters as the subjects of your sentences and their actions as the verbs. By *characters*, I mean tangible, concrete nouns, like sandstone, aspen trees, or T cells. The more concrete the characters and the more vigorous their actions, the better the story. Here is a good example. (In this as well as many other examples, I have underlined the subject in each sentence once and the verb twice.)

In the Great Lakes of Africa, large and diverse species flocks of cichlid fish have evolved rapidly. Lake Victoria, the largest of these lakes, had until recently at least 500 species of haplochromine cichlids. They were ecologically so diverse that they utilized almost all resources available to freshwater fishes in general, despite having evolved in perhaps as little as 12,400 years and from a single ancestral species. This species flock is the most notable example of vertebrate explosive evolution known today. Many of its species have vanished within two decades, which can only partly be explained by predation by the introduced Nile perch (*Lates* spp.). Stenotopic rock-dwelling cichlids, of which there are more than 200 species, are rarely eaten by Nile perch. Yet, many such species have disappeared in the past 10 years.

Notice that the authors use concrete nouns as subjects like *flocks*, *species*, and *cichlids*, which in most sentences do things such as *evolve*, *vanish*, and *disappear*.

We scientists have fascinating stories to tell, but our writing seldom conveys them in a way that is interesting and easy to understand. Rather than choosing concrete characters as subjects, many scientific writers choose abstract nouns. Abstract nouns come from verbs and sometimes adjectives. They name intangible things such as ideas, emotions, or qualities, and they don't work well as characters. Here are some common abstract nouns and the verbs and adjectives they come from:

Verb	Abstract Noun
understand	understanding
observe	observation
interpret	interpretation
assume	assumption
predict	prediction
manipulate	manipulation
demonstrate	demonstration
develop	development

exclude	exclusion
respond	response

Adjective	Abstract Noun
efficient	efficiency
~~accurate~~	~~accuracy~~
applicable	applicability

Abstract nouns are good for one purpose; they help us say concisely what would otherwise take many words to explain. For example, *evolution*, *facilitation*, and *mutation* are well established and widely accepted abstract nouns that condense larger ideas into one word. However, many scientific writers use abstract nouns to make their writing sound more sophisticated, at least to the *writer*. But to readers, abstract nouns are bewildering, especially when they play important roles like subjects. Subjects name the characters of your story. If the subjects in your sentences are abstract instead of concrete, readers have trouble visualizing them, and they end up in a daze.

Furthermore, abstract nouns that act as subjects tend to nudge the characters in a sentence into supporting roles such as modifiers and objects of prepositions where readers are likely to miss them. Here is an example.

The behavioral manifestations of stress responses have been shown to vary greatly between individuals in rodents, pigs, birds, fish, and humans. (21 words)

The subject of this sentence is *manifestations*, an abstract noun that comes from the verb *manifest*. Readers can't connect with such an abstract concept. But some flesh-and-blood characters appear at the end of the sentence, as the objects of the preposition *in*: *rodents, pigs, birds, fish, and humans.* If we transform these characters into subjects, our readers are more likely to be interested.

The verb in this sentence is *have been shown*. It does a poor job of describing what the characters do. However, several verbs lurk in the sentence, disguised as other parts of speech: *behave* (disguised as *behavioral* acts as an adjective), *respond* (disguised as *responses* acts as an object of a preposition), and *vary* (disguised as the infinitive *to vary* acts as an object). Which verb best describes the action is for the author to decide, but obviously one of the things these characters are doing is *behaving*. This concept is expressed, not as a verb, but as the adjective *behavioral*, which modifies the abstract noun *manifestations*. To revise, change *behavioral* to the verb *behave*, and our characters are suddenly doing something readers can follow.

> **Individual <u>rodents, pigs, birds, fish, and humans</u> <u>behave</u> very differently in response to stress. (14 words)**

This revision has concrete characters as subjects: *rodents, pigs, birds, fish, and humans*. The verb *behave* tells the reader what those characters are doing. The revised sentence is shorter, but more importantly, the story is more compelling and easier to understand.

In science, a character can take many forms, from the microscopic to the cosmic.

> **Because of its proximity to the Milky Way, <u>NGC 205</u> presents us with an ideal opportunity to study at high linear resolution the distribution of gas in a dwarf elliptical and to investigate the effects of an interaction on the gas content and star formation history.**

The subject of this sentence is NGC 205, a companion galaxy to Andromeda and a fine example of a concrete character despite being almost three million light years away from Earth.

You can revise many sentences that have abstract nouns as subjects by substituting the scientist(s) themselves, as in the following example.

> **The <u>accumulation</u> of data sets from across the northern hemisphere <u>has enabled</u> us to address both the utility and cause of C and N isotope differences in ECM and SAP fungi. (31 words)**

The subject of the sentence is *accumulation*, an abstract noun derived from the verb *accumulate*. Readers understandably wonder: *who* is doing the accumulating? *Us*, the object of the verb *has enabled*, provides a clue. The sentence contains three other abstract nouns that also cloud its meaning: *utility*, *cause*, and *differences*. To make the sentence clearer, insert a new subject, *we*, that refers to the scientists and change the abstract noun *accumulation* to its verbal form, *accumulated*. In addition, you can get rid of two abstract nouns, *utility* and *cause*, by substituting *why*. You can also return the abstract noun *differences* to the verb form *differ*. The revision is shorter, more concrete, and easier to understand.

We analyzed data sets accumulated from across the northern hemisphere to address why C and N isotopes differ in ECM and SAP fungi. (23 words)

EXERCISE 1

In each sentence, underline the subject once and the verb twice. Is the subject abstract or concrete? Rewrite the sentences choosing concrete nouns as subjects and making their actions the verbs. Then look at the Exercise Key in Appendix 2.

1. Processes undertaken by diverse plants and animals are responsible for such ecological actions as nutrient cycling, carbon storage, and atmospheric regulation.
2. Declines in birth rates have been observed in many developed countries, and demographers expect that the transition to a stable population will eventually occur in many undeveloped nations as well.
3. Variations in magmatism during rifting have been attributed to variations in mantle temperature, rifting velocity or duration, active upwelling, or small-scale convection.
4. The inability of lateral variations in mantle temperature and composition, alone, to account for our observations leads us to propose that another influence was melt focusing.
5. The ability of mudrock seals to prevent CO_2 leakage is a major concern for geological storage of anthropogenic CO_2.

Use Strong Verbs

When we change verbs into abstract nouns, we rob our sentences of strong verbs. Strong verbs enliven a reader's interest by making vigorous connections between the characters in a sentence and the things they act upon. We often substitute weak verbs that describe our characters' actions poorly or not at all. Two verbs are particularly weak: *be* and *have*. Of course, both verbs play important parts in forming certain tenses, and forms of *be* (for example: *is, are, was, were*) are essential in writing definitions like the one below.

The National Science Foundation and the National Institutes of Health are two federal agencies that fund biological research.

But unless absolutely necessary, avoid using forms of *be* and *have* as main verbs. Their appearance often indicates that stronger verbs are hiding somewhere in your sentence, disguised as abstract nouns.

Understanding seasonal habitat ranges and their distribution is critical for Greater Prairie Chicken conservation and management. (16 words)

The subject of this sentence is the verbal phrase *Understanding seasonal habitat ranges and their distribution*. *Understanding* (an abstract noun), comes from the verb *understand*, and *distribution* (another abstract noun), comes from the verb *distribute*. The objects of the preposition *for* are *conservation* and *management*, two abstract nouns from the verbs *conserve* and *manage*. Because all the strong verbs have been converted into nouns, the author is left with *is* for the main verb. Conversely, strong verbs (like *understand, distribute, conserve,* and *manage*) could clearly communicate the character's actions and give this sentence vitality. To revise, introduce a concrete character as the subject and convert the abstract nouns back into strong verbs. The result is a crisper, more direct sentence.

Before we can conserve and manage Greater Prairie Chickens, we must understand their seasonal habitats. (15 words)

Now, the independent clause starting with *we must understand* has a concrete subject *we*, which the reader can identify as a character, and a strong verb *must understand*. In the dependent clause *Before we can conserve and manage Greater Prairie Chickens*, the abstract nouns *conservation* and *management* have been changed into strong verbs *can conserve* and *manage* with the same concrete subject, *we*.

The next example shows how the author uses strong verbs to tell a good story.

> **Among the cells that bear innate immune or germline-encoded recognition receptors are macrophages, dendritic cells (DCs), mast cells, neutrophils, eosinophils, and the so-called NK cells. These cells can become activated during an inflammatory response, which is virtually always a sign of infection with a pathogenic microbe. Such cells rapidly differentiate into short-lived effector cells whose main role is to get rid of the infection; in this they mainly succeed without recourse to adaptive immunity.**

The first sentence is a definition and uses *are*, a form of *be*, as the main verb. The rest of the sentences use strong verbs: *activate*, *differentiate*, and *succeed*. These verbs, in large part, make this writing interesting and engaging.

EXERCISE 2

In each sentence, underline the main verb twice. Revise the sentences by substituting strong verbs for weak verbs. Replace abstract nouns where you can. Then look at the Exercise Key in Appendix 2.

1. Photographs from space taken by satellites are indicators of urbanization and just one of the demonstrations of the human footprint.
2. Weather variables (precipitation, temperature, and wind speed) are key factors in limiting summer habitat availability.
3. A risk management ranking system is the central mechanism for which prioritization of terrestrial invasive species is based.
4. It is clear that Prairie Chickens are closely associated with sagebrush habitat throughout the year.
5. The occurrence of freezing and thawing is an important control on cohesive bank erosion in the region.

Place Subjects and Verbs Close Together

Not only is it important to choose the right verb, it is also important to put the verb in the right place. Once readers identify the subject as the character in the story, they immediately look for the verb that describes what that character is doing. The closer the verb is to the subject, the clearer the sentence. In this well-written example, the authors follow subjects closely with verbs.

> **Horns <u>form</u> during the larval period, from clusters of epidermal cells that detach from the larval cuticle and undergo a local burst of growth. In *Onthophagus taurus*, *O. nigriventris* and *Xylotrupes gideon*, the <u>horns</u> <u>delay</u> growth until very late in the larval period. As animals purge their guts in preparation for metamorphosis, these epidermal <u>cells</u> <u>begin</u> a rapid burst of proliferation and <u>form</u> evaginated discs of densely folded tissue (horn discs) that unfurl to their full length when the animal sheds its larval cuticle and molts into a pupa.**

In each of the three sentences, the subjects are immediately followed by verbs: *Horns form*, *horns delay*, and *cells begin* and *form*. Readers know exactly who the characters are and what they are doing.

On the other hand, if readers have to plow through more than six or seven words past the subject to find the verb, they often forget what the subject was and have to go back. They also tend to ignore those intervening words. If you have more than six or seven words between the subject and the verb, revise to reduce the interruption.

> **<u>Part</u> of our evidence establishing that the p65 product was derived from uncleaved FAT1 and not from the further proteolytic processing of the cleaved FAT1 heterodimer <u>was obtained</u> by the use of the furin-defective LoVo cells. (36 words)**

Twenty-five words lie between the subject of this sentence, *Part*, and the verb *was obtained*. Three of these words are taken up by the prepositional phrase *of our evidence*, and the rest make up the dependent clause modifying the subject, *that the p65 product was derived from uncleaved FAT1 and not from the further proteolytic processing of the cleaved FAT1 heterodimer*. With such a long interruption between the subject and the verb, readers will tend to ignore these words until they find out what the *Part* did, or in this case, what was done to it. A revision that won't make them wait so long could be:

We established that the p65 product was not derived from the further proteolytic processing of cleaved FAT1 heterodimer. Instead, by using furin-defective LoVo cells, we discovered that p65 was derived from uncleaved FAT1. (33 words)

This revision split the example into two sentences, each having a concrete subject (*we*) followed immediately by strong verbs, *established* and *discovered*. The revision is shorter, but more importantly, it clarifies the difference between two possible derivations of p65.

EXERCISE 3

In each sentence, underline the subject once and the verb twice. Revise by placing the subject and the verb close together. Replace abstract nouns where you can. Then look at the Exercise Key in Appendix 2.

1. Environmentally sensitive solutions to the problems associated with continued population growth and development will require an environmentally literate citizenry.

2. Partnerships between professional teachers, scientists, nonprofessional science educators, and administrators are needed to improve the content and effectiveness of science education, particularly in rural areas.

3. Our ability to predict the spatial spread of exotic species and their transformation of natural communities is still developing.

4. The amount of magmatism that accompanies the extension and rupture of the continental lithosphere varies dramatically at rifts and margins around the world.

5. The migration of melts vertically to the top of the melting region and then laterally along the base of the extended continental lithosphere would focus melts toward the eastern part of the basin.

6. Pre-treatment of tenocytes with different concentrations of wortmannin (1, 10, and 20 nM) for 1 h, treated with curcumin (5 µM) for

4 h, and then treated with IL-1β for 1 h, inhibited the IL-1β-induced NF-κB activation.

REFERENCE

1. Hsu, J. The secrets of storytelling. *Scientific American Mind*, August, 46–51 (2008).

CHAPTER 4

Favor the Active Voice

Voice describes whether the subject of the sentence is *doing* the action or *receiving the action. When the subject of a sentence does* the action, the verb is in active voice. When the subject *receives* the action, the verb is in passive voice.

ACTIVE VOICE

The biologist <u>counted</u> the caribou. (5 words)

In this sentence, the subject is *biologist*, and the verb is *counted*; the biologist does the counting, so the verb is active.

PASSIVE VOICE

The caribou <u>were counted</u> by the biologist. (7 words)

In this sentence, the subject is *caribou*, and the verb is *were counted*; the caribou received the counting, so the verb is passive.

Benefits of Active Voice

Using active voice almost always improves your writing. Sentences in active voice reflect the way we speak to each other every day, so readers find them easy to follow. Active sentences also use fewer words. Notice that the active sentence above is two words shorter than the passive sentence. Active sentences have a direct character-action-goal order. Passive sentences reverse this order and have a *be* verb preceding a form of the main verb that often ends in *-ed* or *-en*. Many passive sentences also have a *by* phrase, which explains who is doing the action. These extra words add up; a whole document made up of passive sentences can be 30% longer than one made up of active sentences. It's interesting that many scientific papers submitted for publication come back with the editorial command: "Cut by one-third." Changing many of the verbs from passive to active may be one way to do it! Here is an example.

This hypothesis is supported by the observation that the timing of spring runoff is significantly different between natural and modified basins (Moore et al. 2011). (passive, 25 words)

In this sentence, the subject *hypothesis* receives the action of supporting, thus the verb *is supported* is passive. If you make the verb active, and substitute Moore et al. (2011) as the subject, you can cut out the *be* verb (*is* in this case) and reduce the prepositional phrase *by the observations* to *observing*. The revision is clearer, more direct, and three words shorter.

Moore et al. (2011) support this hypothesis, observing that the timing of spring runoff is significantly different between natural and modified basins. (active, 22 words)

In addition to being shorter and more direct, active sentences force you to name the characters of your stories. In passive sentences, these characters can go unnamed. By omitting characters, you violate the first principle of telling a story: make characters your subjects. In addition to being dull, writing without characters is abstract and lacks persuasive power.

Dramatic improvements in policy and technology are needed to reconfigure agriculture and land use to gracefully meet global demand for both food and biofuel feedstocks. (passive, 25 words)

The subject of this sentence, *improvements*, is an abstract noun, and the verb *are needed* is passive. This structure lulls readers into accepting that no one is doing or receiving the action. A critical reader might ask, "Who needs these improvements?" and, more importantly, "Who should reconfigure agriculture and land use so that the global demand for food and biofuel are met?" The characters that should take responsibility for these actions are missing, and the readers who might act on this advice are confused.

In a revision, let's assume the authors are referring to the United States. Find a suitable character as the subject and a strong, active verb, and the result will be a more direct, personal statement.

The Department of Agriculture <u>must help</u> farmers with new legislation and technology to meet global demand for biofuels without jeopardizing our food supply or environment. (active, 25 words)

This revision has the same number of words, but with the addition of concrete characters, *The Department of Agriculture* and *farmers*, and the active verb, *help*, the writing comes alive, and the task becomes urgent and compelling.

Sometimes, when writers combine passive voice with abstract nouns, the result can be truly impenetrable.

The <u>variation</u> in survivorship referred to as density-dependent mortality <u>has</u> also <u>been related</u> to negative plant-soil biota feedbacks described for a temperate (Parker and Clay 2000; Parker and Clay 2002) and tropical tree species (Hood et al. 2004). (passive, 37 words)

Reading this sentence is like entering a fog. The abstract noun *variation* is the subject of the passive, weak verb *has been related*. Three more abstract nouns, *survivorship*, *mortality*, and *feedbacks* add to the haze. Eight words separate the subject and the verb. All the concrete nouns play minor roles: *plant*, *soil*, and *biota* modify *feedbacks* forming a long, incomprehensible string, *plant-soil biota feedbacks*, and *tree* modifies *species*, which is the object of the preposition *for*. The reader may well ask, "Who are the characters in this story, and what are they doing?"

To revise, use the scientists as concrete subjects and find stronger, active verbs. If your audience is familiar with common ecological terms, it may not be necessary to define density-dependent mortality as the variation in survivorship. Also, try to break up the string *negative plant-soil biota feedbacks*; however, without more information about what the authors mean, the revision is still unclear.

> Parker and Clay (2000, 2002) <u>found</u> that density dependent mortality in a tropical tree species was related to negative feedbacks between plants and soil biota. Hood et al. (2004) <u>found</u> a similar relationship in a temperate tree species. (active, 37 words)

Readers now know who the characters are, and what they are doing. Concrete subjects are closely followed by active verbs, and even though we have two sentences instead of one, the revision is the same length as the original.

Proper Uses of Passive Voice

Although active voice generally clarifies your writing, the passive voice exists for several good reasons.

First, the passive helps you keep the same or similar subjects in a series of sentences in a paragraph (we'll return to this in Chapter 7). If a series of sentences in a paragraph have consistent subjects, it makes for easier reading and comprehension.

> <u>Supernovae</u> <u>deposit</u> enormous amounts of energy into their surroundings. <u>They</u> <u>play</u> a key role in the heating of their host galaxies and in the enrichment of the interstellar medium with heavy elements that form the building blocks of life. <u>They</u> <u>have been</u> <u>well</u> <u>studied</u> at radio, X-ray, infrared, and optical wavelengths, yet the actual explosion mechanism is not well understood.

Here, the author maintains a consistent string of subjects in all three sentences with *Supernovae*, *They*, and *They*. The verbs in the first two sentences are active, but to keep the subject consistent in the third sentence, the author uses the passive verb, *have been studied*.

Second, passive voice helps you move words to strategic parts of a sentence to give them emphasis, or to connect them to words in the preceding sentence. (In this, and many other examples, I use bold in the text to illustrate the principle.)

> The fundamental <u>constant</u> regulating all microscopic electronic phenomena, from atomic physics to quantum electrodynamics,

is the fine-structure constant α. Experimentally, the current value $\alpha = 1/(137.03602 \pm 0.00021)$ is one of the best determined numbers in physics. Theoretically, the reason why nature selects this particular numerical value has remained a mystery, and has provoked much interesting **speculation**. The **speculations** may be divided roughly into three general types.

Here, the author uses active verbs in the first three sentences: *is, is, has remained,* and *has provoked.* At the end of the third sentence, he introduces the word *speculation.* In order to continue seamlessly into the next sentence using *speculations* as the new subject, he uses the passive verb *may be divided.*

Finally, the passive voice can help you compose a sentence where the action that was done is important, but who did it is not. This situation arises frequently when scientists discuss their procedures or methods, where it is implicit that the scientists themselves are doing the action, and to mention them as the subject in every sentence is repetitive and wordy.

ACTIVE VOICE

I cooled the samples on ice, returned them to Arizona State University, and froze them until I used them. (19 words)

PASSIVE VOICE

Samples were cooled on ice, returned to Arizona State University, and frozen until used. (14 words)

Passive voice works well here. It can save words, particularly in a list of procedures where one item is being manipulated in several different ways. However, don't overuse the passive; apply it sparingly and for a reason. Here is a good example of a mix of active and passive sentences from a clear, concise Methods section.

The number of animals in each chamber was then counted in one of two ways. For liberated embryos, we directly counted all embryos in each vial. For egg capsules, we either (1) directly counted embryos in all capsules and summed them for each vial, or (2) counted embryos in 20 haphazardly chosen capsules from the same mass, calculated the mean number of embryos per capsule, and multiplied the number of capsules in a given vial by this average. The latter method was used for temperature experiments because time constraints precluded immediate counts of embryos in each egg capsule.

The authors use passive verbs in the first and fourth sentences, *was counted* and *was used*, respectively. In the fourth sentence, the passive allows the authors to introduce a new subject, *method*, that refers to information at the end of the previous sentence. The second and third sentences are active. They each begin with prepositional phrases: *For liberated embryos* and *For egg capsules*, which is a nice alternative to beginning each sentence with *We*. Avoiding the overuse of *we* may also be the reason the passive was used in the first sentence.

The main point to remember is this: be thoughtful when choosing between active and passive voice. If a good reason exists to use the passive, then use it. Otherwise, use active voice. Your readers will thank you, and your writing will be clearer and more concise.

EXERCISE 4

Read each sentence and underline the subject and double underline the verb(s). Is the verb active or passive? Check your answers in the Exercise Key in Appendix 2.

1. Four big brown bats served as subjects in these experiments, two males and two females.
2. The animals were collected from private homes in Maryland and were housed in the University of Maryland bat vivarium.
3. Bats were maintained at 80% of their *ad lib* feeding weight and were normally fed mealworms only during experiments.
4. We exposed the bats to a reversed 12h dark: 12h light cycle, and we gave them free access to water.

Change the following sentences from passive to active. I've given you the word counts in the originals. How many words are in your revisions? Check your answers with the Exercise Key in Appendix 2.

1. For effective storage of industrial CO_2, retention times of $\sim 10^4$ yr or greater are required. (15 words)

2. It is hypothesized that groundwater pH must have been, on average, highest shortly before the Late Ordovician to Silurian proliferation of root-forming land plants. (24 words)

3. We were compelled to rely on the SOC90 data as no further information on the occupational situation (employed vs. self-employed) or on the size of the firm was available in retrospective form. (32 words)

4. Moreover, it has been demonstrated that mineral-water reactions increase the pH of groundwater even in the presence of abundant acid-producing lichens (Schatz, 1963). (24 words)

CHAPTER 5

Choose Your Words with Care

Scientists often express themselves in long, complicated words. Believe it or not, one reason for this dates back to the 11th century. At that time, people living in what is now England spoke Anglo-Saxon (also known as Old English). In 1066, England was invaded by the Norman French, who defeated the English army at the Battle of Hastings—an event that would have profound effects on the English language. After the invasion, the Normans took control of institutional, religious, and scholarly affairs, and they conducted their business first in Latin and later in Norman French. Over time, their Latin and French words became anglicized and embedded in the lexicon of these institutions. While common people still spoke Anglo-Saxon, those who were educated learned a vocabulary full of words with Latin and French roots—words that were almost always longer and more formal than their Anglo-Saxon counterparts.

During the Renaissance, the English language was expanded again when English scholars who were translating Greek and Latin texts had to anglicize many words that had no English equivalents.

As a result, English is one of the most flexible and varied of all modern European languages; however, it also has a two-tiered vocabulary. When we speak among ourselves at home or with friends, 80% of the words we use are derived from Anglo-Saxon—a language of mostly short words with broad applications. (Look up the meaning of *time*, and you'll see what I mean.) These include the articles *the, a, this, that*; the prepositions *in, on, of, by, with*; the most common verbs and nouns *do, have, make, head, hand, mother, father, sun, man, woman*, and many others.

However, if we wish to participate in a profession such as science, we must learn a different vocabulary we inherited from the Normans and Renaissance scholars peppered with three times as many long, complicated French-, Latin-, and Greek-based words. Words such as anthropogenic, *interpretation*, attribution, and *demonstration* are not rooted in our daily experience or our readers'. Yet once we learn them, who among us can resist padding our writing with these words to sound smarter? From a reader's point of view, this has led to some very bad writing, writing that is needlessly hard to read and understand. Certainly, this vocabulary is useful, but many of us have forgotten how clear and direct simple words can be.

Use Short Words Instead of Long Ones

As Winston Churchill said, "Short words are best, and old words when short, are best of all."[1] By "old words" he meant Old English words, which he used in his writing and speaking to change the course of 20th-century history. If you want to influence your reader with your ideas, resist the temptation to use long Latin- or French-based words where shorter ones will do. Your message will be clearer and have much more impact. Here are some common Latin- and French-based words used in scientific writing and their shorter substitutes.

Long Words	Short Words
implement	put
adhere	stick
develop	make
retain	keep
utilize	use
terminate	end
ascertain	find
facilitate	help
endeavor	try
transmit	send
initiate	start
alteration	change
investigations	work
prescription	plan
subsequent	next

heterogeneous	patchy
spatial	in space
temporal	in time

The following example is typical of most scientific writing.

Although investigations of medieval plague victims have identified *Yersinia pestis* as the putative etiologic agent of the pandemic, methodological <u>limitations</u> <u>have prevented</u> large-scale genomic investigations to evaluate changes in the pathogen's virulence over time. (34 words)

The writers say three things here: *Yersinia pestis* likely caused the Black Death; not much suitable *Y. pestis* DNA is still around; and thus, large-scale genomic studies of its virulence are hard to do. To make this sentence even more difficult, the writers use long, complicated French- and Latin-based words like *investigations, identified, etiologic, methodological, limitations,* and *evaluate.* Fourteen of the words in this sentence are three syllables or more. A simpler, clearer version uses only five.

By studying medieval plague victims, <u>we</u> <u>know</u> that *Yersinia pestis* likely caused the Black Death; however, we don't know how the pathogen's virulence changed over time, because large-scale genomic studies are hard to do. (34 words)

In the revision, I replaced many of the long, complicated words with shorter ones like *know, likely, caused, how, hard,* and *do.* You may see another improvement—I replaced the abstract subject *limitations* with the concrete noun *we.*

In this excerpt from a legendary paper in physics, the writer uses short, familiar words to describe a complex subject: the physics of organisms moving in viscous liquids. Notice how clear the writing is.

There is a very funny thing about motion at low Reynolds number, which is the following. One special kind of swimming motion is what I call a reciprocal motion. That is to say, I change my body into a certain shape and then I go back to the original

shape by going through the sequence in reverse. At low Reynolds number, everything reverses just fine. Time, in fact, makes no difference—only configuration. If I change quickly or slowly, the pattern of motion is exactly the same.

Of the 87 words in this example, 65% are one syllable: the nouns *shape*, *thing*, and *time*; the verbs *be*, *change*, *say*, *go*, and *make*, the prepositions *of*, *at*, *by*, and *in*, and the pronouns *which*, *what*, and *that*. Most of these words come from Old English. We use them when we speak to each other every day, and unless there is good reason not to, we should write the same way.

EXERCISE 6

How many syllables are in each of the underlined words below? Can you predict whether they have Latin, French, or Old English roots? Look up the origins of each of the words, and then rewrite the sentences using as many Old English words as you can without changing the meaning. List the number of syllables and roots of the shorter words. Check your answers in the Exercise Key in Appendix 2.

1. For example, <u>expansion</u> of the extent of the winter range by <u>continued</u> <u>pioneering</u> of segments of the northern Yellowstone elk herd northward from the park <u>boundary</u> and <u>extensive</u> use of these more northerly areas by greater numbers of elk have been <u>coincident</u> with <u>acquisition</u> and <u>conversion</u> of rangelands from livestock <u>production</u> to elk winter range.

2. We conclude that snag <u>retention</u> at <u>multiple</u> <u>spatial</u> and <u>temporal</u> scales in recent burns, which will be salvage-logged, is a <u>prescription</u> that must be <u>implemented</u> to meet the <u>principles</u> of <u>sustainable</u> forest <u>management</u> and the <u>maintenance</u> of biodiversity in the boreal forest.

Keep Terms the Same

Many scientific writers believe that repeating the same term for an important character makes their writing boring or repetitive. So instead they use different terms. By doing so they risk confusing their readers who think they mean different things. For a reader, consistent terms are the opposite of boring; they are essential to navigating new, complex information.

> In relatively unproductive ecosystems like deserts, **grazers** and **predators** are so rare as to be negligible, and competition for resources structures **plant communities**. In more productive systems like grasslands, a large effective **herbivore community** can be supported and grazing determines **plant biomass**. (42 words)

Here the writer compares two habitats, deserts and grasslands, and describes what limits plants in each. Comparisons are hard for readers because they must remember two or more situations while mentally weighing one against another. Anything that muddies the comparison, like different terms for the same thing, makes it harder. Here, the terms for characters are different: *plant communities* in the first sentence and *plant biomass* in the second. Does the writer mean the same thing? *Herbivores* are also important characters, but they come in two guises as well: *grazers* and *herbivore community*. *Predators* may also be important, but after their first mention, they disappear. Are they crucial to the story? Although most readers can muddle through writing like this, the author has done them no favors. Their concentration is constantly interrupted while they sort out different terms. To revise, refer to the main characters using consistent terms.

> In relatively unproductive ecosystems like deserts, **plant biomass** is limited by a lack of resources. In more productive systems like grasslands, **plant biomass** is limited by herbivores. (27 words)

Now readers can compare the two habitats easily. The main character in both is the same: *plant biomass*, which is limited in deserts by scarce resources and in grasslands by herbivores. If predators are also important, they can be explained in a separate sentence.

Find the different terms in the following sentences and underline them. Revise by replacing different terms with ones that are the same. Use as many short, simple words as you can. Check your answers in the Exercise Key in Appendix 2.

1. One way to assess the perceived risk of feeding in different locations is to measure the proportion of the available food a forager removes before switching to an alternative patch. All else being equal, foragers should be willing to forage longer and remove more food from a safe area than a risky one.

2. Stress coping styles have been characterized as a proactive/reactive dichotomy in laboratory and domesticated animals. In this study, we examined the prevalence of proactive/reactive stress coping styles in wild-caught short-tailed mice (Scotinomys teguina). We compared stress responses to spontaneous singing, a social and reproductive behavior that characterizes this species.

3. Antimicrobial resistance genes allow a microorganism to expand its ecological niche, allowing its proliferation in the presence of certain noxious compounds. From this standpoint, it is not surprising that antibiotic resistance genes are associated with highly mobile genetic elements, because the benefit to a microorganism derived from antibiotic resistance is transient, owing to the temporal and spatial heterogeneity of antibiotic-bearing environments.

4. Studies of long-term outcomes in offspring exposed to maternal undernutrition and stress caused by the Dutch Hunger Winter of 1944 to 1945 revealed an increased prevalence of metabolic disease, such as glucose intolerance, obesity, and cardiovascular disease, as well as emotional and psychiatric disorders. Animal models have been developed to assess the long-term consequences of a variety of maternal challenges including under- and overnutrition, hyperglycemia, chronic stress, and inflammation. Exposures to a wide range of insults during gestation are associated with convergent effects on fetal growth, neurodevelopment, and metabolism.

Break Up Noun Strings

Scientific writers have another bad habit: putting nouns and often their modifiers together into long unwieldy strings. Noun strings are difficult for readers because they don't know which word in the string is important. Familiarity is the key here as well as length. Short, familiar strings are often useful; they let us name complex concepts like *population dynamics* (and *noun string*) in very few words. If your readers are familiar with the string—and it saves words—use it, but avoid long strings you have rarely seen before.

> As the **labor market time commitment** of mothers has increased in western societies in the recent decades, <u>questions</u> about the provisions of care for children, especially in relation to maintaining and generating time for care, <u>have attracted</u> **significant international social and policy attention.** (43 words)

This sentence starts out with a four-word noun string, *labor market time commitment*, which is the subject of the dependent clause starting with *As*. A reader might reasonably wonder, which noun best describes the relevant aspect of *mothers*; is it mothers' *labor*, mothers' *time*, or mothers' *commitment*? Other problems abound: the subject of the independent clause *questions* is abstract and is separated from the verb *have attracted* by 17 words. The object *attention* is encumbered with four modifiers: *significant, international, social,* and *policy*. To revise, break up the noun string to emphasize the aspect of a working mother's life—time—that is most relevant to the discussion. Replace the abstract subject with a concrete noun and put it close to its verb. Then, rearrange the modifiers of *attention* to make a clearer statement.

> As mothers in western societies commit more time to work, <u>they spend</u> less time with their children. This <u>phenomenon has attracted</u> widespread attention from social scientists and policy makers. (29 words)

Here, I changed the abstract noun *commitment* to the verb *commit*, and broke up the rest of the noun string *labor market time* with the help of the preposition *to*: *as mothers . . . commit more time to work*. Now, the concrete subject *they* is followed immediately by the verb *spend*, and the modifiers of *attention* have been broken up with the help of the preposition *from*. Although the revision has two sentences, it is 14 words shorter than the original.

EXERCISE 8

In the following sentences, underline the noun strings and revise by breaking them up and replacing long words with short ones. Check your answers in the Exercise Key in Appendix 2.

1. Developing regular exercise programs and diet regimes contributes to disease risk prevention and optimal health promotion.

2. Research focused on care time deficits and time squeezes for families has identified the persistence of gendered care time burdens and the sense of time pressure many dual-earner families experience around care.

3. There will be major conservation implications if mercury ingestion in ospreys causes negative population level effects either through direct mortality or negative fecundity.

Rethink Technical Terms

By technical terms, I mean the in-house language of any group of people who share a particular expertise or interest. Technical terms are one of the hallmarks of science, and as scientific fields become more specialized, we invent new terms to describe them. These terms let the in-group express complex ideas briefly and efficiently; however, for the out-group, these terms are incomprehensible. One immunologist I spoke with said he was having trouble understanding the writing of colleagues in subdisciplines within his own field!

Just as real estate brokers summarize what is most important about a property in the mantra "location, location, location," a scientific writer should repeat the mantra "audience, audience, audience." Use specialized technical terms only when you are sure that all your readers will understand them. Some technical terms are familiar to almost everyone and don't need explanation, DNA and 3-D for example, but if there is any doubt about what your audience knows, be conservative and define your technical terms or leave them out.

Aneuploidy and translocations lead to progressive alterations in chromosome structure and epigenetic modifications characteristic of tumorigenesis. (16 words)

This example contains technical terms from the field of genetics: *Aneuploidy, translocations, epigenetic modifications,* and *tumorigenesis.* Only readers familiar with advanced genetics will understand them. However, for the reader with a limited background in genetics, this sentence is incomprehensible. For a broader audience, a translation that would minimize the technical terms might be:

Relative to the cell lines from which they arose, cells prone to become tumors characteristically show abnormal chromosome numbers, chromosomal rearrangements, and aberrant patterns of gene expression arising from defects in gene regulation. (33 words)

This revision shows why technical terms are so useful to those who understand them. The revision contains over twice as many words as the original, and this translation still includes *chromosomal rearrangement, gene expression,* and *gene regulation,* which might take several additional sentences to explain. Technical terms obviously save words, but if you are writing to an audience who may not understand them, define them or leave them out. Here is another example.

Several entities have petitioned for Prairie Chicken listing consideration under the ESA, and the USFWS responded to this with a positive 90-day finding in 2005. (25 words)

The terms *ESA,* the *USFWS,* and *a positive 90-day finding* are familiar to those who understand the legal complexities of the Endangered Species Act. For someone outside that group, the meaning of this sentence won't be clear. In a revision, explain to the reader what the terms mean or leave them out.

Several groups have petitioned the United States Fish and Wildlife Service (USFWS) to list Prairie Chickens under the Endangered Species Act (ESA). The USFWS has responded within the mandated period of 90 days that listing is warranted. (37 words)

Here, the acronyms USFWS and ESA are defined, and the legal term, *a positive 90-day finding*, is explained. Once this is done, the writer can use them without confusing her readers.

The next example is a well-written introductory paragraph from a journal with a broad scientific audience. The writers define terms well, so that readers from other backgrounds can make sense of the ideas.

Many biological systems have evolved to work with a very high energetic efficiency. . . . At first glance, the beating of cilia and flagella does not fall into the category of processes with such a high efficiency. Cilia are hair-like protrusions that beat in an asymmetric fashion to pump the fluid in the direction of their effective stroke. They propel certain protozoa, such as *Paramecium*, and also fulfill a number of functions in mammals including mucous clearance from airways, left-right asymmetry determination, and transport of an egg cell in fallopian tubes.

By defining cilia and explaining their functions, these writers prepare their readers for the rest of the paper, which introduces a measure for energetic efficiency in beating cilia and calculates their optimal beating pattern.

EXERCISE 9

Here is an example of complex writing from the field of immunology. Have the authors tried to make it understandable to a broad scientific audience? Have they succeeded? Why or why not? Check your answers in the Exercise Key in Appendix 2.

1. One of the well-researched immunoregulatory functions of probiotics is the induction of cytokine production. In particular, the induction of IL-10 and IL-12 production by probiotics has been studied intensively, because the balance of IL-10/IL-12 secreted by

macrophages and dendritic cells in response to microbes is crucial for determination of the direction of the immune response. IL-10 is an anti-inflammatory cytokine and is expected to improve chronic inflammation, such as that of inflammatory bowel disease and autoimmune disease. IL-10 downregulates phagocytic and T cell functions, including the production of proinflammatory cytokines, such as IL-12, TNF-α, and IFN-γ, that control inflammatory responses. IL-10 promotes the development of regulatory T cells for the control of excessive immune responses. In contrast, IL-12 is an important mediator of cell-mediated immunity and is expected to augment the natural immune defense against infections and cancers. IL-12 stimulates T cells to secrete IFN-γ, promotes Th1 cell development, and, directly or indirectly, augments the cytotoxic activity of NK cells and macrophages. IL-12 also suppresses redundant Th2 cell responses for the control of allergy.

REFERENCE

1. Churchill, W. An elder statesman as man of letters. *New York Times Magazine*, Nov. 13, 78–79 (1949).

Scientific writing is often wordy. We have all faced the daunting task of reading a complicated paper made worse by verbose writing. Our goal as writers should be to express our ideas with no surplus words, omitting what our readers can easily infer.

One advantage of using plain English is that it immediately makes your writing more concise. Notice that most of the revisions we have done already are shorter than the originals. The example below shows how wordy writing becomes more concise when you apply the principles we've discussed so far.

> **Inhalation of vapor phase particulate matter chemical contaminants from biomass combustion in domestic settings is a significant contributor to local disease burden. (22 words)**

This sentence has several problems. The subject *Inhalation* is an abstract noun derived from the verb *inhale*. It is followed by three prepositional phrases: *of vapor phase particulate matter chemical contaminants* (which contains a string of six nouns), *from biomass combustion*, and *in domestic settings*. Thirteen words separate the subject from the verb. The verb *is* is weak. Strong verbs that might give this sentence vigor have been changed into their noun equivalents: *inhale* into *inhalation*, *combust* into *combustion*, and *contribute* into *contributor*.

The writer also used the most complicated words she could think of to describe some very simple concepts: *vapor phase particulate matter chemical contaminants* means smoke, *biomass combustion* means wood burning, *is a significant contributor* means causes, and *disease burden* means health problems. By applying the principles of plain English, we get a clear, concise sentence.

> **Domestic wood smoke causes local health problems. (7 words)**

Smoke is a concrete noun. *Causes* is a strong verb. The strings of nouns have been broken up and replaced with words readers can understand.

The subject and the verb are close together. The revised sentence has seven words compared to the original 22—one-third as long!

Redundancy

Most needless words that inflate scientific writing are redundant— words that can be cut with no harm to the message. Redundancy comes in many forms; several of the most common are described below.

1. REPETITION

Instead of making one clear statement, writers often repeat themselves using slightly different words. This wastes words and hides the main point.

Despite the widely recognized importance of instream wood and organic debris dams in forested stream ecosystems, analytical approaches to quantify the spatial extent and pattern of instream wood distribution are rare and the usefulness of available metrics has been seldom evaluated. Wood influences stream geomorphology, biotic habitat, and biogeochemical cycling, therefore quantifying the spatial distribution of instream wood is important for understanding the corresponding distribution of key stream functions. (69 words)

A list of the main points in this paragraph might look like this:

1. Wood in streams is important.
2. Ways to measure wood in streams are few and seldom evaluated.
3. Because it affects stream functions, wood in streams is important.
4. Because it affects stream functions, measuring wood in streams is important.

Now it's easier to see the repetition. One main point—that wood in streams is important—is repeated, and another—that measuring wood in streams is important—becomes self-evident. With a little reorganizing and simplifying, you can eliminate the repetition. The result is clear and concise.

Pieces of wood and the dams they produce influence stream geomorphology, biotic habitat, and biogeochemical cycling. Despite this, there are few methods for measuring wood in streams, and these are seldom evaluated. (32 words)

2. EXCESS DETAIL

It's also easy to give readers too much detail.

Nature is ripe with examples of the exquisite mechanisms organisms utilize to salvage survivorship in the face of unfortunate circumstances, and regeneration is one such process. Regeneration is a way that stick insects cope with shedding a limb to survive either fouled molt or a predation attempt, but like other organisms, this process often comes with associated costs. (58 words)

In the first sentence of this Abstract, the writer could describe regeneration more concisely without references to *nature*, *exquisite mechanisms*, and *unfortunate circumstances*. In the second sentence, the comparison with other organisms is also unnecessary, unless the writer will elaborate on this later. A revision should state the main points without the excess detail.

Many organisms regrow lost limbs through a process called regeneration. In stick insects, regeneration allows individuals to survive a predation attempt or a fouled molt, but at a cost. (29 words)

Clearly, how much detail you provide depends on what your readers know and the purpose of your writing. For instance, a student may include much more about a topic's background in a class paper than his professor would in a scholarly paper. Each writes for a different audience: the student for the teacher who will assess his knowledge, and the professor for an informed audience of peers.

3. A WORD FOR A PHRASE

Scientific writers habitually use wordy phrases that could be expressed more concisely. These phrases are so common that some journal editors keep lists of them. The editor of a prestigious journal contributed his list of wordy phrases and their shorter substitutes to the list below.[1]

Wordy Phrase	Shorter Substitute
in this study we assessed	we assessed
conduct an investigation of	investigate
were responsible for	caused
played the role of	were
in order to	to
for the following reasons	because
during the course of; during the process of	during
a majority; most of the	most
undertake an examination of	study
various lines of evidence	evidence
the analysis presented in this paper	our analysis
in the absence of	without
located in; located at	in; at
in the vicinity of; in close proximity to	near
in no case; on no occasion	never
at the present time; at this point in time	now
an example of this is the fact that	for example

Wordy phrases are easy to cut from your writing—it's just a question of recognizing them.

4. "THE"

You can frequently omit the article "the" from your text without any loss of meaning. One example comes from an earlier chapter.

Certainly, the merit of your scientific writing rests as much on the content as on the style. Equally important are the questions, the hypotheses, the experimental designs, and the interpretations you describe. (32 words)

Deleting many of the articles produces a shorter, crisper sentence.

Certainly, the merit of your scientific writing rests as much on content as on style. Equally important are the questions, hypotheses, experimental designs, and interpretations you describe. (27 words)

Metadiscourse and Transition Words

Writing about your thinking or your writing is known as *metadiscourse*. We use metadiscourse to good effect in almost everything we write: to comment on our ideas—*I believe, to summarize, in conclusion*; to structure what we write—*first, second, more importantly*; and to guide our readers through our writing—*note that, in order to understand, consider now*.

You may recognize useful metadiscourse by another name—transition words. Transition words are like road signs; they help readers navigate a written piece by making smooth connections between sentences and paragraphs. Used skillfully, they can make reading almost effortless. The following paragraph shows how transition words can be used to good purpose.

Dorsal, which is activated by the Toll pathway during dorso-ventral axis formation, does not appear to play a role in the systemic immune response in adult flies. **Instead**, another NF-κB family member—Dif (drosophila immunity factor)—is required for the induction of Drosomycin by Toll. **Additionally**, Spatzle is required for the activation of Toll by fungal pathogens; **however**, the serine protease cascade that generates active Spatzle during development is not involved in the immune response. **Therefore**, a different protease cascade must regulate its processing.

Here the author used words such as *instead, additionally, however,* and *therefore* to direct readers along a complicated developmental pathway.

While you need some metadiscourse in almost everything you write, too much can bury your ideas. Knowing the difference between too much and just enough is tricky, especially for writers new to a field.

In this essay I will be talking about how forest fragmentation causes declines in neotropical migrants. **The particular**

articles that I read for this essay really gave me a good idea of the background of this issue. It is well known that neotropical migrants are declining because of deforestation in their wintering habitat, **but what was new to me was learning that** forest fragmentation in eastern North America was also detrimental. (71 words)

In the introduction to her paper, this writer pads her ideas with a good deal of metadiscourse: *In this essay I will be talking about how, The particular articles I read for this essay really gave me a good idea of the background of this issue, It is well known that, and but what was new to me was learning that.* Some of this should be cut, but some of it could stay. For instance, her professor may be interested in learning how the information she gathered changed her views. Leaving in some metadiscourse can also create a more conversational tone.

It is well known that neotropical migrants are declining because of deforestation in their wintering habitat, **but what was new to me was learning that** forest fragmentation in eastern North America was also detrimental. (34 words)

In the first two sentences, I removed needless metadiscourse and repetitive information (no need to say that forest fragmentation is causing declines in neotropical migrants twice). However, I kept the metadiscourse in the third sentence to create a conversational tone and to highlight what the student learned.

Another kind of needless metadiscourse leads your reader through the text.

In the previous section of this paper, I concluded that the problem of rising sea levels was important. **In this next section**, I would like to describe the additional problem of ocean acidification. (33 words)

This becomes:

The next problem is ocean acidification. (6 words)

Some metadiscourse places an unspecified observer in the text who has seen or found something. In a revision, just state what was observed.

High birth rates **have been observed** to occur in parts of the Midwest that **have been determined** to have especially high rates of unemployment. (24 words)

This becomes:

High birth rates occur in parts of the Midwest that have especially high rates of unemployment. (16 words)

Other sources of metadiscourse are hedges and emphatics. These are words we use to convey caution or confidence. When overdone, they can contribute to wordiness and give your reader the wrong impression of you as a writer and a scientist.

Hedges are words such as *usually, often, sometimes, perhaps, may, might, can, could, seem,* and *suggest.* Although hedges often convey real uncertainty, when we use them needlessly and often, we sound unsure.

We found that juveniles **were more likely to move** towards the speaker, approaching closer and more quickly, during the simulated song interactions than during solo song or control playback trials. These results **indicate** that juveniles are especially interested in eavesdropping on song contests and **suggest** that these types of social interactions **may be** particularly powerful tutoring events.

The verbs, *were more likely to move, indicate, suggest,* and *may be,* give the reader the impression that the study's results were inconclusive. However, in this case, the results strongly supported the hypothesis, so there was no need to sound timid.

We found that juveniles **were particularly attracted** to countersinging interactions and approached playbacks of these song interactions **significantly more** than simulated solo singing or the control playback trials. This result **is consistent** with the social eavesdropping hypothesis that juveniles **may learn** to sing via eavesdropping on the singing interactions of adults.

Here, the verbs and their modifiers are stronger and more definitive: *were particularly attracted*, *significantly more*, and *is consistent*. This writing is more likely to convince a reader that the results are conclusive.

Another reason to use hedges with care is that it is easy to overly qualify something. Look again at the last sentence in the original example above.

These results **indicate** that juveniles are especially interested in eavesdropping on song contests and **suggest** that these types of social interactions **may be** particularly powerful tutoring events.

This sentence holds three qualifiers: *indicate*, *suggest*, and *may be*. There is no need to include so many when the same level of uncertainty could be conveyed with just one: *suggest*.

These results **suggest** that juveniles are especially interested in eavesdropping on song contests and that these types of social interactions are particularly powerful tutoring events.

Emphatic words are the opposite of hedges. They are used for emphasis and include words like *clearly*, *very*, *obviously*, *indeed*, *undoubtedly*, *certainly*, *major*, *primary*, and *essential*. Although they don't appear often enough in scientific writing, it's refreshing when they do. Because they highlight important information, emphatics help broad audiences understand unfamiliar topics. Emphatics are also important in competitive writing such as grant proposals, as in this example from Chapter 2.

Horned beetles provide an **unusual** opportunity to combine studies of trait development with experiments exploring sexual selection and the evolutionary significance of enlarged male weapons (horns). By building on almost ten years of research directed towards this goal, the PI now has the opportunity to forge a **truly integrative** research program, offering **unique** possibilities for inspiring and training young scientists, and providing a **comprehensive** picture of the evolution of some of nature's **most bizarre** animal shapes.

Emphatic words and phrases such as *unusual, truly integrative, unique, comprehensive,* and *most bizarre* helped convince the reviewers that this project was worth funding.

Affirmatives and Negatives

Affirmative statements are often less wordy than negative statements. In an affirmative statement, the subject is *doing* something to something else, while in a negative statement, the subject is *not doing* something to something else. Negative statements are more opaque, because they imply what should be happening by stating what is not. Affirmatives state the point more directly. Obviously, in sentences where you are contradicting or denying some point, only the negative will do, but re-phrasing many negatives to affirmatives will save words. The list below gives some examples.

Negative	Affirmative
did not accept	rejected
did not consider	ignored
does not have	lacks
did not allow	prevented
not the same	different
not possible	impossible
not many	few

In the following example, the writers use negative statements to make their points.

> The canopy cover of Norway maples **does not allow** sufficient light to penetrate the understory and native tree seedlings **cannot germinate.** (21 words)

If you change the two negatives, *does not allow* and *cannot germinate* to the equivalent affirmative statements, your writing is more direct, and you will save a word.

The canopy cover of Norway maples **blocks** sufficient light from penetrating the understory and **prevents** native tree seedlings from germinating. (20 words)

Because words find so many ways to multiply, concise scientific writing is rare and notable indeed. One famous example is the single-page paper on the structure of DNA by J. D. Watson and F. H. C. Crick excerpted below.[2]

We wish to put forward a radically different structure for the salt of deoxyribose nucleic acid. This structure has two helical chains each coiled round the same axis (see diagram). We have made the usual chemical assumptions, namely, that each chain consists of phosphate diester groups joining β-D-deoxyribofuranose residues with 3', 5' linkages. The two chains (but not their bases) are related by a dyad perpendicular to the fibre axis. Both chains follow right-handed helices, but owing to the dyad the sequences of the atoms in the two chains run in opposite directions.

EXERCISE 10

Describe the causes of wordiness in the following sentences and paragraphs where needed. Rewrite omitting as many needless words as you can. I've given you the word counts of the originals. How many words are in your revisions? Check your answers in the Exercise Key in Appendix 2.

1. While a growing body of research indicates that large herbivores as a group can exert strong indirect effects on co-occurring species, there are comparatively few examples of strong community-wide impacts from individual large herbivore species. (37 words)

2. Small mammal species diversity increased in exclosures relative to controls, while survivorship showed no significant trends. (16 words)

3. In this essay, I will be looking at how higher summer temperatures cause quicker soil and plant evaporation. We all know that climate change has caused elevated temperatures in the Northwest throughout the spring and summer months. We also know that these record-

breaking temperatures have the effect of quickly and easily desiccating soil and drying out plant foliage so that it is more flammable. Understandably then, when lightning strikes this very combustible environment, a spark can very quickly turn into a widespread blaze. (83 words)

4. Zimbabwean undocumented migrants are shown to be marginalized and vulnerable with limited transnational citizenship. (14 words)

5. When the lithosphere extends and rifts along continental margins, magma is produced in varying quantities. Widely spaced geophysical transects show that rifting along some continental margins can transition from magma-poor to magma-rich. Our wide-angle seismic data from the Black Sea provide the first direct observations of such a transition. This transition coincides with a transform fault and is abrupt, occurring over only ~20–30 km. This abrupt transition cannot be explained solely by gradual along-margin variations in mantle properties, since these would be expected to result in a smooth transition from magma-poor to magma-rich rifting over hundreds of kilometers. We suggest that the abruptness of the transition results from the 3-D migration of magma into areas of greater extension during rifting, a phenomenon that has been observed in active rift environments such as mid-ocean ridges. (133 words)

6. The empirical data presented in this article reveal a segmented labor market and exploitation, with undocumented migrants not benefiting from international protection, human rights, nation state citizenship rights, or rights associated with the more recent concepts of postnational and transnational citizenship. (41 words)

List the transition words in the following paragraph and explain why they are useful. Check your answer in the Exercise Key.

7. The systemic immune response in *Drosophila* is mediated by a battery of antimicrobial peptides produced largely by the fat body, an insect organ analogous to the mammalian liver. These peptides lyse microorganisms by forming pores in their cell walls. Functionally, the antimicrobial peptides fall into three classes depending on the pathogen specificity of their lytic activity. Thus, Drosomycin is a ma-

jor antifungal peptide, whereas Diptericin is active against gram-negative bacteria, and Defensin works against gram-positive bacteria. Interestingly, infection of Drosophila with different classes of pathogens leads to preferential induction of the appropriate group of antimicrobial peptides.

The next paragraph is about hybrids and their long-term survival and reproduction. How many different principles of plain English can you identify in the writing? Describe them. Is the writing concise as a result? Check your answer in the Exercise Key.

8. Introgressive hybridization is most commonly observed in zones of geographical contact between otherwise allopatric taxa. Studies of such zones have provided important insights into the evolutionary process and have helped resolve part of the debate about fitness of hybrids. In many cases, most hybrid genotypes tend to be less fit than are the parental genotypes in parental habitats, owing either to endogenous or exogenous selection or both. However, theory predicts that some can be of equal or superior fitness in new habitats and, occasionally, even in parental habitats.

REFERENCE
1. Moore, R. *Writing to Learn Biology* (Saunders College Publishing, 1992).
2. Watson, J. D. & Crick, F. H. C. Molecular Structure of nucleic acids. *Nature* **171**, 737–738 (1953).

CHAPTER 7

Old Information and New Information

So far, I've discussed principles of plain English that will help you write a clear, concise sentence—one with a concrete subject, a strong, active verb, and few unnecessary words. But we don't write sentences in isolation; we write them in sequences, and these sequences must dovetail to form sensible paragraphs. This requires thinking about sentences in a whole new way. In fact, sentences have definite beginnings and ends, and readers expect to find specific information in each. If you place the right information at the beginnings and ends of your sentences, you will meet your readers' expectations, and your paragraphs will form cohesive units.

Put Old Information at Beginnings of Sentences

Organize each paragraph around one character or idea, and make it the subject of each of your sentences. These subjects become familiar to your readers and are called the old information. Because the subject precedes the verb in most English-language sentences, you will be grounding your readers in familiar territory where they need it most—at the beginnings of your sentences. For instance, if your paragraph is about linear models, then make linear models, or words that mean the same thing, your subjects. Each time your readers see one of these subjects, they will recognize it and know what you're talking about. This allows them to concentrate on what you tell them *about* those subjects. Here is a good example.

> **Quantum mechanics** has enjoyed many successes since its formulation in the early 20th century. **It** has explained the structure and interactions of atoms, nuclei, and subnuclear particles, and has given rise to revolutionary technologies, such as integrated circuit chips and magnetic resonance imaging. At the same time, **it** has generated puzzles that persist to this day.

The writers have used consistent subjects at the beginnings of their sentences to refer to the same character: *Quantum mechanics*, *It*, and *it*. As you can see, keeping your subjects consistent doesn't mean using the same word for them. As long as your reader can make the connections between them easily, you're fine.

Put New Information at Ends of Sentences

With old, familiar information at the beginnings of your sentences followed by strong verbs, the ends of the sentences are free to take the new information your reader hasn't heard yet. Like the punch line of a good joke, this is the most interesting part of your story. Readers pay particular attention to information at the ends of sentences, so save them for what you want your reader to remember. Look again at the last example.

> **Quantum mechanics has enjoyed many successes since its formulation in the early 20th century. It has explained the structure and interactions of atoms, nuclei, and subnuclear particles, and has given rise to revolutionary technologies, such as integrated circuit chips and magnetic resonance imaging. At the same time, it has generated puzzles that persist to this day.**

The writers have followed their subjects closely with strong, active verbs: *has enjoyed, has explained,* and *has generated.* The second sentence is broken into two clauses, each with its own verb, *has explained* and *has given rise.* The ends of all the sentences (and in the second sentence, the clauses) hold the new information *about* quantum mechanics: its successes, its uses, and its puzzles.

Obviously, not all paragraphs have consistent subjects. In the example above, the writers may have felt that quantum mechanics was familiar to their readers, so they start their paragraph (an introduction to an essay), without explanation. However, if your character is likely to be unfamiliar to your readers, you should devote one or two sentences to describe it—remembering to introduce your character where readers are most likely to notice it—at the ends of those descriptive sentences. Here is a good example from Chapter 3.

> **In the Great Lakes of Africa, large and diverse species flocks of cichlid fish** have evolved **rapidly. Lake Victoria, the largest of these lakes, had until recently at least 500 species of haplochromine cichlids. They** were **ecologically so diverse that they utilized almost all resources available to freshwater fishes in general, despite having evolved in perhaps as little as 12,400 years and from a single ancestral species. This species flock is the most notable example of vertebrate explosive evolution known today. Many of its species** have vanished **within two decades, which can only partly be explained by predation by the introduced Nile perch (Lates spp.). Stenotopic rock-dwelling cichlids, of which there are more than 200 species, are rarely eaten by Nile perch. Yet, many such species** have disappeared **in the past 10 years.**

These writers introduce the main characters, *cichlid fish*, by placing them at or near the ends of the first two sentences. The first sentence begins with the prepositional phrase *In the Great Lakes of Africa. Flocks* is the subject, but it, and the prepositional phrase, *of cichlid fish*, occur near the end of the sentence. The subject *Lake Victoria* begins the next sentence, and *haplochromine cichlids*, the object of the preposition *of*, occurs at the end. Once the reader is well aware of them, *cichlids* or similar terms (*they, flock, many [of its species], cichlids,* and *species*) are subjects at the beginnings of the remaining sentences.

In addition to introducing an unfamiliar character, there will be other times when you must switch subjects in a paragraph. Use the same technique. Introduce the new character at the end of a sentence, where the reader will notice it, then, in the next sentence, use the new character as the subject. *Never* surprise readers by introducing a new character at the beginning of a sentence. First alert them to the new character at the end of the previous sentence. Here is a good example.

Supernovae deposit enormous amounts of energy into their surroundings. **They** play a key role in the heating of their host galaxies and in the enrichment of the interstellar medium with heavy elements that form the building blocks of life. Yet, the actual explosion **mechanism** is not well understood. One **way** to study the explosion is through the dynamics of the stellar debris that comprise supernova remnants such as **Cassiopeia A. Cas A** is the 2nd youngest known supernova remnant in the Galaxy (approximately 340 years old) and is also among the brightest. **It** is well studied at radio, X-ray, infrared, and optical wavelengths and is known to have two oppositely directed jets of ejecta with expansion velocities as high as 15,000 km/s.

Early in the paragraph, *supernovae* is the main character and it serves as the subject at the beginning of each of the first four sentences: *Supernovae, They, mechanism* (modified by *explosion*), and *way* (to *study the explosion*). At the end of the fourth sentence, we are introduced to another character, *Cassiopeia A*, so we are prepared when it becomes the new subject of the next two sentences: *Cas A* and *It*.

Misplacing old and new information is a common error in scientific writing. The results are writing that lacks "flow," and readers who miss the main points. Here are several ways these errors occur:

1. Writers often shift old information to the end of the sentence, which nudges the new information to the beginning. As a result, the new information, which holds the main point, is not where readers expect it. Only the writer can decide what points he wants to make, but whatever they are, they must be placed at the ends of sentences. The following example highlights this problem.

Wildland **fires** are disturbances that occur with long recurrence intervals and generally high severity in some forest types and with shorter intervals and lower severity in others. For millennia, wildland **fires** have arguably been the most important disturbance process throughout many western forests. Seed germination and establishment, growth patterns, plant community composition and structure, rates of mortality, soil productivity, and other properties and processes of western forest ecosystems are often strongly influenced and shaped by **fire disturbance regimes**. Even so, perhaps the most controversial aspect of western land management at present is the ecology of **fire and fire management**.

In each of the first two sentences, the main character *fires* is the subject. This old information is followed by new information about that subject: it occurs at long and short intervals, and it is the most important disturbance in many western forests. The third sentence abruptly places new information, *germination and establishment* and so on, at the beginning, and the old information, *fire disturbance regimes*, at the end. The fourth sentence does the same. The new information, that fires are controversial, is near the beginning of the sentence, while the old information, *fire and fire management*, is at the end. One revision keeps fire as the consistent subject near the beginnings of the sentences and places the new information about why it is important, and that it is controversial, at the ends.

For millennia, wildland **fires** have been arguably the most important disturbances in many western forests. In some forest types, these **fires** occur infrequently with generally high severity, while in others, they occur frequently with lower severity. **Fire** strongly influences many aspects of western forest ecosystems including germination and establishment of seedlings, patterns of growth, composition, and structure of plant communities, rates of mortality, and productivity of soils. Even so, of the issues western land managers face, the ecology of **fire** and **fire management** are the most controversial.

2. Another error occurs when writers repeat the old information at the ends of their sentences.

Riparian forests in western North America are exceptionally important habitats and their ecological significance is often disproportionately important in relation to the amount of landscape they occupy. The productivity of **riparian habitats** is typically much higher than adjacent areas, and many species of plants and animals are restricted to **riparian habitats**. In areas of the arid west, **riparian forests** constitute less that 1% of the landscape, and yet well over 50% of the species of breeding birds depend on **those habitats**.

Besides changing his terms, this writer ends the last two sentences with the old information, *riparian habitats* and *those habitats*. This shifts the new information to the middle of the sentences where readers don't notice it. Revise by trimming the old information from the ends of these sentences.

Riparian forests in western North America are exceptionally important. Although they occupy only a small part of the landscape, **they** are much more productive than adjacent areas, and many species of plants and animals are restricted to them. For example, in the arid west, **riparian forests** constitute less than 1% of the landscape, yet they support well over 50% of the breeding bird species.

3. Sometimes writers put the new information at the beginning of an introductory sentence instead of the end.

There are different **ways** in which international migrants can gain protection and/or rights. The first is the protection that exists for refugees under the terms of the 1951 Geneva Convention.

In the first sentence, the information that will be developed in the next several sentences is *ways*. However, *ways* occurs at the beginning of the sentence, as the subject. Put *ways* at the end of the first sentence, and you will alert your reader that a list will follow.

International migrants can gain protection and/or rights in different **ways**. The first is the protection that exists for refugees under the terms of the 1951 Geneva Convention.

4. Sometimes writers add unnecessary words to the end of a sentence. These words hide the new information.

During my elementary school years, I found that my most exciting and rewarding moments came from the sense of wonder that I felt after learning something completely new to me for the first time. A second wave of amazement often arrived days or weeks later, when I actually began to understand what I thought I had learned.

This writer has added unnecessary words at the ends of both sentences, *to me for the first time*, means the same thing as *new*, and *what I thought I had learned* could be said more concisely. Deleting these words will make the main points stand out.

During my elementary school years, I found that my most exciting and rewarding moments came after learning something completely new. A second wave of amazement often arrived days or weeks later, when I actually began to understand it.

Note that I used the pronoun *it* at the end of the revised second sentence. Pronouns that replace unnecessary words or phrases at the ends of sentences put extra emphasis on the preceding word, in this case, *understand*.

EXERCISE 11

In the following sentences and paragraphs, did the writers put the old and new information in the right places? Revise any incorrect sentences using the principles we've discussed. Then look at the Exercise Key in Appendix 2.

1. Unfortunately, as noted 40 years ago, few students experience the thrill of doing field science because they are rarely allowed to leave the confines of the classroom to become immersed in field-based science.
2. Bank erosion rates along the South River in Virginia increased by factors of 2–3 after 1957. Increased bank erosion rates cannot be explained by changes in the intensity of either freeze-thaw or storm

intensity, and changes in the density of riparian trees should have decreased erosion rates.

3. Students majoring in science often believe they can escape the intensive writing and presentations that their peers in the humanities and social sciences must do. However, science is a collective human endeavor whose success hinges upon effective communication, both written and oral. Even if findings are ground breaking, they are potentially worthless if they can't be shared with others in a clear and engaging way. Teaching undergraduate science students to effectively communicate is therefore an essential goal.

4. Climate plays an important part in determining the average numbers of a species, and periodical seasons of extreme cold or drought, I believe to be the most effective of all checks. I estimated that the winter of 1854–55 destroyed four-fifths of the birds in my own grounds; and this is a tremendous destruction, when we remember that ten per cent is an extraordinarily severe mortality from epidemics with man. The action of climate seems at first sight to be quite independent of the struggle for existence; but in so far as climate chiefly acts in reducing food, it brings on the most severe struggle between the individuals, whether of the same or of distinct species, which subsist on the same kind of food. Even when climate, for instance extreme cold, acts directly, it will be the least vigorous, or those which have got least food through the advancing winter, which will suffer most.

CHAPTER **8**   **Make Lists Parallel**

In a single sentence, you often tell your readers more than one piece of new information. Two or more pieces of information are easier to read and remember if their structures are parallel. Parallel structure refers to the way items in any kind of list are written using similar kinds of words within similar grammatical arrangements. Here is a good example.

> A successful phenomenology must accomplish many things: **it must explain** why repetitions of the same measurement lead to definite, but differing, outcomes, **and** why the probability distribution of outcomes is given by the Born rule; **it must permit** quantum coherence to be maintained for atomic and mesoscopic systems, **while** predicting definite outcomes for measurements with realistic apparatus sizes in realistic measurement times; **it should conserve** overall probability, **so** that particles do not spontaneously disappear; and **it should not allow** superluminal transmission of signals.

If you dissect the items in the list above, you see how each is described in a similar way:

it must explain . . . , and
it must permit . . . , while
it should conserve . . . , so
and it should not allow . . .

Each item is described in an independent clause with the consistent subject it followed by one of two auxiliary words, *must* or *should*, followed by an active verb, *explain, permit, conserve,* or *allow.* The first three independent clauses contain dependent clauses beginning with the conjunctions *and, while,* and *so.* The last clause lacks a dependent clause, but the rhythmic quality of the repeated sequences is maintained.

The following list is not parallel.

These similarities include **an early sensitive period, an innate filtering mechanism that isolates conspecific vocalizations, a babbling developmental phase,** and **the importance of social variables in vocal learning.**

This sentence lists four phases of song learning in birds. Each is described in a single phrase, but the phrases differ substantially, making the list difficult to read and remember. The items in the list are:

an early sensitive period,
an innate filtering mechanism that isolates conspecific
    vocalizations,
a babbling development phase, and
the importance of social variables in vocal learning.

The first phrase begins with the article *an* followed by two adjectives, *early* and *sensitive*, followed by the noun *period*. The pattern repeats in the third phrase, *a babbling developmental phase*, but changes in the second and fourth. The second phrase begins the same way, *an innate filtering mechanism*, but includes a dependent clause, *that isolates conspecific vocalizations*, and the fourth phrase begins with a different article, *the* followed by an abstract noun *importance* and two prepositional phrases, *of social variables* and *in vocal learning*.

To make the list parallel, structure each phrase the same way. I deleted the dependent clause from the second phrase, *that isolates conspecific vocalizations*, because presumably the writer will elaborate later. Also, I picked one term to describe the main character. In the original, there are three: *period*, *mechanism*, and *phase*. The revised list looks like this:

an early sensitive phase,
a filtering phase,
a babbling phase, and
a social phase.

Now, all the items are parallel.

These similarities include **an early sensitive phase, a filtering phase, a babbling phase,** and **a social phase.**

Dissect the lists in the following sentences. Are they parallel? Revise them using the principles we've discussed. Then check your answers with the Exercise Key in Appendix 2.

1. Central to this deficit has been the rising average age of the nursing workforce and the decline in the number of hours worked; fewer nurses are working standard full-time hours (35–44 hours per week) and 44 percent work part-time.

2. The problem of finding the optimal strokes of hypothetical micro-swimmers has drawn a lot of attention in recent years. Problems that have been solved include the optimal stroke pattern of Purcell's three-link swimmer, an ideal elastic flagellum, a shape-changing body, a two- and a three-sphere swimmer, and a spherical squirmer.

3. Cilia are hair-like protrusions that beat in an asymmetric fashion to pump the fluid in the direction of their effective stroke. They propel certain protozoa, such as *Paramecium*, and also fulfill a number of functions in mammals, including mucous clearance from airways, left–right asymmetry determination, and transport of an egg cell in fallopian tubes.

4. Integrons consist of three elements: an attachment site where the horizontally acquired sequence is integrated; a gene encoding a site-specific recombinase (that is, integrase); and a promoter that drives expression of the incorporated sequence.

5. North American (NA)-EEEV strains cause periodic outbreaks of mosquito-borne encephalitis in humans and equines, are highly neurovirulent, and, in comparison with related Venezuelan equine encephalitis virus (VEEV) and western equine encephalitis virus (WEEV), cause far more severe encephalitic disease in humans.

CHAPTER 9

Vary the Length of Your Sentences

The best writing consists of sentences of various lengths. A string of long sentences (30 words or more) is difficult to get through; a string of short sentences (10 words or less) is choppy, and a string of medium-length sentences (15–25 words) is monotonous. Scientific writers can hardly be accused of writing too many short sentences; instead, they tend to write medium to long sentences and to maintain the same sentence length throughout a whole document. Here is an example.

> One of the major <u>goals</u> of conservation biology <u>is</u> to conduct scientific research that will aid in the preservation of natural landscapes. (22 words) Of particular concern to scientists and environmentalists <u>are</u> natural <u>areas</u> that have remained relatively undisturbed for long periods of time. (20 words) These <u>areas</u> often <u>serve</u> as habitats for a variety of plant and animal species that are not found in more disturbed areas. (22 words) Accordingly, these <u>lands</u> <u>are</u> often <u>set</u> aside as protected areas. (10 words) These <u>areas</u>, although protected from urbanization and development, <u>are</u> often subject to high levels of disturbance from recreational activities. (19 words) Thus, land <u>managers</u> <u>must struggle</u> to find an acceptable balance between biological and social management objectives. (16 words)

This writing is praiseworthy in many respects: the subjects are concrete; the verbs are mostly active (although due to a plethora of abstract nouns, many are weak); the subjects and verbs are close together, and the old and new information is placed correctly. However, almost all the sentences are similar lengths (16–22 words), and the resulting monotony can be hypnotic. If you vary the length of the sentences by combining some and shortening others, you're more likely to hold your readers' attention.

Conservation biologists strive to preserve natural landscapes. (7 words) They are particularly concerned with areas that have a long history of protection where a variety of plants and animals can be found that are absent from more disturbed areas. (30 words) Often however, these areas are subject to high levels of disturbance from recreational activities. (14 words) By providing quantitative data on the effects of recreation on the surrounding biota, conservation biologists can help land mangers find a balance between social and biological demands. (27 words)

In this revision, the sentence lengths vary from seven words (short), to 14 and 27 words (medium), to 30 words (long). The result is more interesting to read.

In the following well-written example, the writers vary the length of their sentences, from 11 to 26 words. The first and last sentences are divided by dashes that give the impression that each is made up of two short sentences: in the first, four and seven words long, and the last, six and 10 words long. Notice how short sentences have punch and serve to emphasize important points, while longer sentences add rhythm. Use both in your writing.

Most nematodes are gonochoristic—they produce *XO* males and *XX* females. (11 words) The males make small, round spermatids that activate following mating, extend pseudopods, and crawl toward the spermathecae of the female, where they compete to fertilize oocytes. (26 words) However, some species have evolved an androdioecious mating system, with *XO* males and self-fertile *XX* hermaphrodites. (16 words) In these species, the hermaphrodites make spermatids late in larval development and then permanently switch to the production of oocytes. (20 words) In almost all respects, the male and hermaphrodite sperm from these species appear identical. (14 words) However, one trait is sexually dimorphic—the male sperm are much larger than those of hermaphrodites. (16 words)

The average sentence length in this paragraph is 18 words, below the recommended maximum of 20–25 words. If you keep your average sentence length below this maximum, you avoid too many long sentences, such as the one below.

> **Further processing [of the heterodimer] to generate an intracellular fragment able to traverse to the nucleus was dependent upon ligand binding, resulting in casein kinase-dependent phosphorylation followed by enzymic cleavage likely involving an ADAM (a disintegrin and metalloprotease) type metalloprotease sequentially followed by further intracytoplasmic cleavage by the γ-secretase complex. (49 words)**

This sentence is far too long, especially considering its complexity. To revise, break up the series of actions, so that most occur in their own sentences. Long sentences typically have more than one point, so breaking them up is usually straightforward.

> **Further processing of the heterodimer to generate an intracellular fragment able to traverse to the nucleus was dependent upon ligand binding, resulting in casein kinase-dependent phosphorylation. (26 words) This was followed by enzymic cleavage likely involving an ADAM (a disintegrin and metalloprotease) type metalloprotease. (16 words) The product was then cleaved by the γ-secretase complex. (12 words)**

When the steps are delivered in shorter sentences, the reader can follow them easily.

EXERCISE 13

Count the number of words in these sentences. Are there a variety of sentence lengths? What is the average number of words in a sentence? Is the average long or short? How would you suggest improving the sentence length? Give an example. Then check your answers in the Exercise Key in Appendix 2.

1. In order to unravel the mode of action of neuronal networks, a neurobiologist's dream would be not only to be able to monitor neu-

ronal activity but also to have control over distinct sets of neurons and to be able to manipulate their activity and observe the effect on behavior. This idea is not new. As the activity of a neuron is based on the depolarization of its cell membrane, neuronal activity can be induced by an experimenter using stimulation electrodes by which the cell membrane can be artificially depolarized or hyperpolarized. Although stimulation electrodes have served, and continue to serve, neuroscientists well for decades, limitations of this invasive approach are obvious.

Break up the following long sentences into shorter ones.

2. The extrapolation from *in vitro* measurements to the *in vivo* behavior of proteins is hampered by extremely high (300–400 mg/mL) intracellular macromolecular concentrations in the cell, i.e. crowding, which is one of the most important factors that influences the structure and function of proteins under physiological conditions. (47 words)

3. Some of the confusion about the role of hybridization in evolutionary diversification stems from the contradiction between a perceived necessity for cessation of gene flow to enable adaptive population differentiation on the one hand, and the potential of hybridization for generating adaptive variation, functional novelty, and new species on the other. (52 words)

CHAPTER 10

Design Your Paragraphs

At this stage, you are ready to shift your focus from sentence-level revision to paragraph design. Paragraphs, like sentences, have parts—an issue, a development, a conclusion, and a point—and readers expect to find specific information in each. As they progress through a paragraph, readers depend on these parts to help them recognize new ideas, organize supporting evidence, and recap. If any parts are missing, or the information is misplaced, readers become disoriented. I'll explain each of the parts in the context of two different kinds of paragraphs: those that begin documents, called introductory paragraphs, and those that make up the rest of the document, called body paragraphs.

First, a few words about length. Imagine your readers as they begin one of your paragraphs. Don't overwhelm them with long blocks of unbroken text. Count the words in your paragraphs and break up the ones with 200 words or more. Ideally paragraphs should be about 150 words long, with some variation from one to the next.

Issue

Each paragraph should open by telling the reader what the paragraph is about. Most of us call this a topic sentence, but the term can be misleading, because good writers often open their paragraphs with two or three sentences. A more accurate term is *issue*. No matter how long it is, the issue should end by introducing the characters and actions that will be featured in the rest of the paragraph. The paragraph below has a one-sentence issue.

> **Determining the cause of HIV-1 resistance in Old World monkey cells stymied HIV researchers for nearly two decades.** An early view was that the block resulted from expression of an incompatible receptor on the surface of Old World

monkey cells. However, identification of the HIV-1 co-receptor in the mid-1990s disproved this hypothesis. Subsequent studies demonstrated that HIV-1 could enter Old World monkey cells, but a block that targeted the viral capsid prevented the establishment of a permanent infection.

The issue is the first sentence, and it summarizes the main idea the writer develops in the rest of the paragraph: how Old World monkeys' resistance to HIV-1 has stumped researchers for two decades. It also introduces the paragraph's main characters and actions: *HIV-1, Old World monkeys, stymied,* and *researchers.*

The issue in the next paragraph is two sentences long.

We initially hypothesized that TRIM5α functioned as a cofactor necessary for capsid uncoating. However, subsequent findings argued against this hypothesis. First, knocking down human TRIM5α showed no effects on HIV-1 replication in human cells. Second, rodent cells, which do not express TRIM5α, supported HIV-1 infection if engineered to express an appropriate receptor. Finally, human TRIM5α does not associate with the HIV-1 capsid in biochemical assays. Thus, TRIM5α appeared to have evolved primarily as an inhibitory factor aimed at thwarting viral replication, rather than a host factor co-opted by HIV-1 to promote infection.

In the first sentence, the writer presents a preliminary hypothesis that in the second sentence he rejects. The rest of the paragraph describes the research that led to the rejection. The characters and actions to be developed in the rest of the paragraph are introduced near the end of the issue: *subsequent findings* and *argued.*

Development
After presenting the issue at the beginning of a paragraph, the writer expands on it in the *development.* One good way to write the development is to lay out well-defined steps that lead to some conclusion, as the writer of the previous example did.

We initially hypothesized that TRIM5α functioned as a cofactor necessary for capsid uncoating. However, subsequent findings argued against this hypothesis. **First, knocking down human TRIM5α showed no effects on HIV-1 replication in human cells. Second, rodent cells, which do not express TRIM5α, supported HIV-1 infection if engineered to express an appropriate receptor. Finally, human TRIM5α does not associate with the HIV-1 capsid in biochemical assays.** Thus, TRIM5α appeared to have evolved primarily as an inhibitory factor aimed at thwarting viral replication, rather than a host factor co-opted by HIV-1 to promote infection.

The development is three sentences long, with each sentence beginning with a transition word that helps the reader navigate the information: First, Second, and Finally.

Other ways to develop an issue include giving examples or presenting expert opinion. Another is qualifying the issue in some way. Sometimes, the issue is a question, and the writer uses the development to answer it.

Does TRIM5α have the ability to block infection by other retroviruses? **We found that TRIM5α from various Old World monkey species conferred potent resistance to HIV-1, but not SIV. New World monkey TRIM5α proteins, in contrast, blocked SIV but not HIV-1 infection. Human TRIM5α inhibited N-MLV and EIAV replication.** Thus the variation among *TRIM5* orthologs accounts for the observed patterns of post-entry blocks to retroviral replication among primate species.

The development answers the question posed in the issue and consists of three sentences describing TRIM5α's ability to block infection in Old World monkeys, New World monkeys, and humans.

Conclusion

Once readers have been introduced to a new idea, and that idea has been developed, they look for a sentence at the end of the paragraph that serves as a kind of comprehension check—did they understand what the writer was getting at? That last sentence, called the *conclusion*, gives them a moment to think about it. Give your readers that moment before introducing them to a new idea in the next paragraph. Here are two adjacent paragraphs as an example.

> Does TRIM5α have the ability to block infection by other retroviruses? We found that TRIM5α from various Old World monkey species conferred potent resistance to HIV-1, but not SIV. New World monkey TRIM5α proteins, in contrast, blocked SIV but not HIV-1 infection. Human TRIM5α inhibited N-MLV and EIAV replication. **Thus the variation among *TRIM5* orthologs accounts for the observed patterns of post-entry blocks to retroviral replication among primate species.**
>
> To determine why Old World monkey TRIM5α, but not human TRIM5α, potently blocks HIV-1, we systematically altered the human sequence to more closely resemble the monkey sequence. Remarkably, we found that a single amino acid determines the antiviral potency of human TRIM5α. If a positively charged arginine residue in the C-terminal domain of human TRIM5α is either deleted or replaced with an uncharged amino acid, human cells gain the ability to inhibit HIV-1 infection. **Perhaps some humans have already acquired this change and are naturally resistant to HIV-1 infection.**

In the first paragraph, the conclusion sums up the answers to the question posed in the issue. Summaries are common types of conclusions. They rephrase the information from the development and put it into a broader context. Now that readers know that TRIM5α blocks infection by other retroviruses, they are ready to tackle a new question in the issue of the second paragraph: why does Old World monkey TRIM5α, but not human TRIM5α, block HIV-1?

The end of the second paragraph in the example shows another way to write a conclusion. Posing a question or making a speculation is a natural extension of the development. This kind of conclusion leaves the reader with something to ponder.

There are many ways to conclude paragraphs, but you should avoid writing the issue that belongs at the beginning of the next paragraph. Paragraph conclusions are places to summarize, speculate, or question your reader on the information that has gone before, not surprise him with the unfamiliar. Save your issues for introducing new paragraphs.

Point

So far, we've discussed the issue, development, and conclusion in a body paragraph; however, you must change this basic design in one important way in an introductory paragraph. We still consider the introduction to a whole document as the issue, but it may take up to three paragraphs or more depending on the length of the document. However long it is, there should be one sentence at the end of the last paragraph that foretells the whole text. Traditionally, this is called a thesis, but to avoid confusing it with a well-used term in science, a better term is point. The point is important because it prepares the reader for what is to come—without it, confusion reigns. Here is the introduction to the prize–winning essay from which I've taken all the examples in this chapter.[1] Look for the point.

Humans have been exposed to retroviruses for millions of years. Indeed, a significant portion of our genome consists of endogenous retroviruses—reminders of our vulnerability to past infections. The HIV/AIDS epidemic, which began nearly a century ago when simian immunodeficiency virus (SIV) passed from chimpanzees into a human host, is the latest episode in the long-standing coevolutionary struggle between retroviruses and their hosts.

Human immunodeficiency virus type 1 (HIV-1) causes AIDS in humans, and to a lesser extent, in chimpanzees. However, not long after the discovery of HIV-1, scientists realized that certain primate species were resistant to HIV-1 infection. In particular, monkeys from Africa and Asia, referred to as Old World monkeys, could not be infected with HIV-1 and did not develop AIDS. This discovery brought both excitement and frustration. The block to

HIV-1 replication in Old World monkey cells hindered efforts to develop an animal model for testing drugs and vaccines. On the other hand, Old World monkeys had evolved for millions of years in Africa—the epicenter of the current HIV-1 epidemic. Perhaps exposure to past HIV-1–like epidemics led to the emergence of an antiviral defense that protects them against HIV-1.

The point, the sentence that foretells what is to come, is at the end of the second paragraph. There we learn about an antiviral defense that protects Old World monkeys from HIV-1 infection—the theme that is developed in the rest of the essay.

EXERCISE 14

Find the issue in the following body paragraphs. How does the writer develop it? Check your answers with the Exercise Key in Appendix 2.

1. Males of *O. acuminatus* [dung beetles] employed two very different tactics to encounter and mate with females: they either attempted to monopolize access to a female by guarding the entrance to her tunnel (guarding), or they attempted to bypass guarding males (sneaking). Guarding behavior entailed remaining inside a tunnel with a female, and fighting intruding males over possession of the tunnel. Guarding males blocked tunnel entrances and periodically "patrolled" the length of the tunnel. Rival males could gain possession of a tunnel only by forcibly evicting the resident male, and both fights and turnovers were frequent. Fights over tunnel occupancy entailed repeated butting, wrestling and pushing of opponents, and fights continued until one of the contestants left the tunnel.

2. One prerequisite for the maintenance of dimorphism is that organisms experience a fitness tradeoff across environments. If animals encounter several discrete environment types, or ecological or behavioral situations, and these different environments favor different morphologies, then distinct morphological alternatives can evolve within a single population—each specialized for one of the different environments. Such fitness tradeoffs have been demonstrated for several dimorphic species. For example, soft and hard seed diets

have favored two divergent bill morphologies within populations of African finches, and high and low levels of predation have favored alternative shell morphologies in barnacles, and spined and spineless morphologies in rotifers and *Daphnia*. It is possible that the alternative reproductive tactics characterized in this study produce a similar situation in *O. acuminatus*. If guarding and sneaking behaviors favor horned and hornless male morphologies, respectively, then the reproductive behavior of males may have contributed to the evolution of male horn length dimorphism in this species.

3. In recent times, the origin of the adaptive immune response has been uncovered. It turns out that the two recombinase-activated genes are encoded in a short stretch of DNA, in opposite orientations and lacking exons. This suggested an origin in a retroposon, as did the presence of the recognition signal sequences that lie 3' of all V gene segments and 5' of all J gene segments. This hypothesis was tested *in vitro* and shown to be true. Other processes expand diversity tremendously, such as the generation of D gene segments in the first chain to rearrange, the nucleotide-adding enzyme TdT that inserts nucleotides in the junctions of V-D-J junctions, and somatic hypermutation.

Describe the issue, development, and conclusion in each of the following paragraphs.

4. Despite a consistent correlation between genome size and the obligate association with host cells, genome reduction is not simply an adaptive response to living within hosts. Instead, the trend toward large-scale gene loss reflects a lack of effective selection for maintaining genes in these specialized microbes. Because the host presents a constant environment rich in metabolic intermediates, some genes are rendered useless by adoption of a strictly symbiotic or pathogenic life-style. These superfluous sequences are eliminated through mutational bias favoring deletions, a process apparently universal in bacterial lineages. Thus, all of the fully sequenced small genomes display a pattern of loss of biosynthetic pathways, such as those for amino acids that can be obtained from the host cytoplasm.

5. Unlike pathogens, symbionts may devote part of their genomes to

processes that are more directly beneficial to the host rather than to the bacterial cell itself. *Buchnera* retains and even amplifies genes for the biosynthesis of amino acids required by hosts, devoting almost 10% of its genome to these pathways, which are missing from pathogens with similarly small genomes. Because of their fastidious growth requirements, the biological role of obligately associated symbionts can rarely be determined experimentally. However, genome comparisons can provide a means for determining their functions in hosts. Such future research should reveal, for example, whether the endosymbionts of blood-feeding hosts, such as *Wigglesworthia glossinia* in tsetse flies, retain pathways for biosynthesis of vitamins absent from blood, whether the symbiont *Vibrio fischeri* provides functions other than bioluminescence to its squid host, and whether the mutualistic *Wolbachia* of filarial nematodes contain genes for host benefit that are absent in the parasitic *Wolbachia* of arthropods.

REFERENCE

1. Stremlau, M. Why Old World Monkeys are resistant to HIV-1. *Science* **318**, 1565–1566 (2007).

CHAPTER 11

Arrange Your Paragraphs

By the time you finish revising the design of your paragraphs, you may need to rearrange some of them. Good writers arrange their paragraphs in predictable patterns that help readers move easily through a document and grasp the main points. Some of these patterns are described below.

Normally, writers use several of these patterns in a single document. In addition, they often use these patterns when arranging sentences within a paragraph, especially when the paragraph contains a lot of information.

Chronological Order

Science is a process—a sequence of related actions that produces a result. For example, an experiment may involve several steps, each occurring in a specific order, or a biochemical pathway may consist of several reactions. The most logical way to present this information is in chronological order. Once the order is determined, keep the order the same throughout the whole document. Here is a good example from the Methods section of a paper on dung beetles.

> To compare the reproductive behaviors of horned and hornless males, I observed their methods of mate-acquisition both without and with competition from a rival male. **In the first experiment**, one male and one female were placed in each of 12 observation chambers. Seven of these males were hornless, five were horned, and females were selected at random. Beetles were observed for a minimum of three half-hour intervals (maximum of six intervals), during which time all behaviors were recorded.
>
> **In the second experiment**, two males (one horned and one hornless) were placed together in each observation chamber with a single female, and behaviors of all individuals were monitored as above. This second experiment tested for behavioral

differences arising as a result of direct competition for access to females. Because dung always had large numbers of *O. acuminatus*, and because horned and hornless males occurred in approximately equal frequencies on Barro Colorado Island, this experiment accurately reflected natural conditions experienced by males.

The writer arranged the paragraphs describing each experiment in chronological order. He followed the same order in the Results section, with a paragraph describing the outcomes of experiment 1, then one describing the outcomes of experiment 2, and so forth. Certainly, scientific investigations are not always linear. Your results may lead you to change your original hypothesis and start again. In such cases, reporting your experiments in chronological order won't work. Instead, organize the experiments so your story is logical.

Arranging paragraphs chronologically usually works well when you describe a process; however, scientific writing consists of many kinds of information that call for other patterns of arrangement.

General to Specific

Every story begins with an introduction. The Introduction section to a scientific paper for instance, provides the background your readers need to understand an unfamiliar topic. A logical way to present unfamiliar information is to progress from general to specific. In the first paragraph, give a general overview of the topic, and in succeeding paragraphs, gradually narrow to the specific focus of the paper. Here is a good example.

Predation is a major cause of mortality for most species of animals, and many produce **alarm calls** when they perceive a potential predator. **Alarm calls** often differ in acoustic structure, depending on the situation in which they are produced. If a species is preyed upon by different predators that use different hunting strategies or vary in the degree of danger they present, selection can favor variation in alarm calls that encode this information. **Such variation in alarm calls** can be used to transfer information about the type of predator, the degree of threat that a predator represents, or both.

In addition to discriminating among broad types of predators (e.g., raptor versus snake), discriminating among morphologically similar predators within a single type (e.g., different species of raptors) could also be adaptive if the predators vary in the degree of threat they pose. One species that is faced with numerous, morphologically similar predators is the **black-capped chickadee** (*Poecile atricapilla*). Chickadees are small, common songbirds that are widespread throughout North America. In the non-breeding season, chickadees form flocks of six to eight birds. **They use an elaborate system of vocalizations** to mediate social interactions in these flocks and **to warn conspecifics about predators.**

In the first paragraph, the authors introduce alarm calls as a way many animals avoid predation. Then they describe different kinds of alarm calls that encode information about different kinds of predators. Near the end of the second paragraph, they introduce one particular animal whose alarm calls may distinguish between similar types of predators—the black-capped chickadee.

The Discussion section of a scientific paper usually adopts the opposite arrangement, progressing from specific findings to general implications. Here is a good example.

In total, our results show that the introduction of foxes to the Aleutian archipelago transformed the islands from grasslands to maritime tundra. Fox predation reduced seabird abundance and distribution, in turn reducing nutrient transport from sea to land. The more nutrient-impoverished ecosystem that resulted favored less productive forbs and shrubs over more productive grasses and sedges.

These findings have several broad implications. First, they show that strong direct effects of introduced predators on their naïve prey can ultimately have dramatic indirect effects on entire ecosystems and that these effects may occur over large areas—in this case across an entire archipelago. Second, they bolster growing evidence that the flow of nutrients, energy, and material from one ecosystem to another can subsidize populations and, importantly, influence the structure of food webs. Finally, they show that the mechanisms by which predators exert ecosystem-level effects extend beyond both the original conceptual model provided by Hairston *et al.* and its more recent elaborations.

In the first paragraph, the authors summarize the specific results of their experiments. The second paragraph is more general and describes the broad implications of those results.

Least Important to Most Important

One purpose of scientific writing is to persuade. Persuasive writing benefits from an arrangement that moves from least important evidence to most important. For the same reason that we place new, important information at the ends of sentences—readers are more likely to remember it—placing important ideas later in your work can be more persuasive. This principle holds true whether the new information is at the end of a sentence, a paragraph, or a group of paragraphs. Here is an example from the introduction to a book chapter subsection.

Many species of crustaceans have enlarged appendages used in fighting. In stomatopods, for example, the second pair of maxillipeds has been lengthened and strengthened to produce powerful weapons, the "raptorial appendages," which are used in both prey capture and fighting (figure 4.5). A subset of the stomatopods, species termed "smashers," use their raptorial appendages to disable armored prey such as mollusks and crabs. When these weapons are used against conspecifics, they are capable of causing serious injury, even death. In another group, the snapping shrimps, the claw (or chela) of one of the first pair of walking legs is greatly enlarged. This claw can be closed rapidly to produce an audible snap. A snap produced against the body of a conspecific is capable of causing severe injury.

Not surprisingly, crustaceans often use these and similar weapons in aggressive displays; the weapons are extended, waved, or otherwise brandished during agonistic encounters. Such displays may function to signal aggressive intentions; in particular, brandishing a weapon may signal that attack is imminent. A weapon display also may serve to signal fighting ability, if weapon size is important in defeating opponents, or if weapon size correlates with body size and body size is important in defeating opponents.

The idea that weapons function to signal fighting ability has also been suggested for other categories of weapons, such as antlers of deer and the horns of sheep, however, the empirical evidence for a signal function of weapons is slim in all these groups. **We believe that the evidence is stronger for crustaceans, which is one reason for concentrating on these animals; the other is that weapon displays in crustaceans provide some of the best evidence of deception available for any type of aggressive signal**.

In the first paragraph, the writers describe several examples of crustacean appendages used as weapons. In the second paragraph, we are told that these weapons can also be used in displays. It isn't until the end of the third paragraph that we are told why crustaceans (rather than animals with other types of weapons) were chosen for this study: there is stronger evidence that crustacean weapons are used as signals of fighting ability, and they afford a unique opportunity to study deception.

Problem to Solution

Another way to arrange paragraphs is to present a problem or paradox early in the document and proceed to solve it or show a novel approach to it. Scientists are continually reassessing and debating the results of previous research, so this type of arrangement is common. Here is a good example.

When applied to ciliary propulsion, Lighthill's efficiency has some drawbacks. For one, it is not a direct criterion for the hydrodynamic efficiency of cilia as it also depends on the size and shape of the whole swimmer. Besides that, it is naturally applicable only for swimmers and not for other systems involving ciliary fluid transport with a variety of functions, like left-right asymmetry determination. **We therefore propose a different criterion for efficiency at the level of a single cilium or a carpet of cilia.** A first thought might be to define it as the volume flow rate of the transported fluid, divided by the dissipated power. However, as the flow rate scales linearly with the velocity, but the dissipation quadratically, this criterion would yield the highest efficiency for infinitesimally slow cilia, just like optimizing the fuel consumption of a road vehicle alone might lead to fitting it with an infinitesimally weak engine. Instead, like engineers trying to optimize the fuel consumption at a given speed, the well-posed question is which beating pattern of a cilium will achieve a certain flow rate with the smallest possible dissipation.

The problem of finding the optimal strokes of hypothetical microswimmers has drawn a lot of attention in recent years. Problems that have been solved include the optimal stroke pattern of Purcell's three-link swimmer, an ideal elastic flagellum, a shape-changing body, a two- and a three-sphere swimmer, and a spherical squirmer. Most recently, Tam and Hosoi optimized the stroke patterns of *Chlamydomonas* flagella. **However, all these studies are still far from the complexity of a ciliary beat with an arbitrary 3D shape, let alone from an infinite field of interacting cilia. In addition, they were all performed for the swimming efficiency of the whole microorganism,** whereas **our goal is to optimize the pumping efficiency at the level of a single cilium,** which can be applicable to a much greater variety of ciliary systems.

These writers pose a series of problems and possible solutions. The problems described in the first paragraph involve the limits of one calculation of the efficiency of ciliary propulsion. In the second paragraph, other problems involve the narrow focus of calculations based on a single microorganism. At the end of the second paragraph, the writers propose a conceptual path toward solving these problems—by understanding the efficiency of single cilium.

Compare and Contrast

In compare and contrast arrangements, writers usually devote one paragraph to the attributes of one character or idea and another paragraph to the attributes of a similar or different character or idea. This ordering helps readers separate the characters while they weigh their similarities or differences. Several paragraphs that describe one character or idea can be grouped together, or the paragraphs can alternate between characters. Here is a good example from the Results section of the beetle paper.

> **Males of O. acuminatus employed two very different tactics to encounter and mate with females:** they either attempted to monopolize access to a female by **guarding** the entrance to her tunnel (guarding), or they attempted to bypass guarding males **(sneaking). Guarding** behavior entailed remaining inside a tunnel with a female, and fighting intruding males over possession of the tunnel. Guarding males blocked tunnel entrances and periodically "patrolled" the length of the tunnel. Rival males could gain possession of a tunnel only by forcibly evicting the resident male, and both fights and turnovers were frequent. Fights over tunnel occupancy entailed repeated butting, wrestling and pushing of opponents, and fights continued until one of the contestants left the tunnel.
>
> **Sneaking** involved bypassing the guarding male. The primary method of sneaking into tunnels was to dig side-tunnels that intercepted guarded tunnels below ground. New tunnels were dug immediately adjacent (< 2 cm) to a guarded tunnel. These tunnels then turned horizontally 1–2 cm below ground, and often intercepted primary tunnels beneath the guarding male (16/24 side-

tunnels). In this fashion, sneaking males sometimes bypassed the guarding male and mated with females undetected (observed in four instances).

In the first paragraph, the writer introduces *guarding* and *sneaking*. The rest of the first paragraph is devoted to describing guarding and the second to sneaking. Arranged in this way, the reader can easily distinguish the differences in the two behaviors.

Clearly, not all paragraphs are arranged in the patterns I've described. Many paragraphs simply elaborate on the topic of the previous paragraph or provide equally weighted lines of evidence supporting a single result. Here is an example from the introduction to an essay in a scientific journal.

> **Bacteria form intimate and quite often mutually beneficial associations with a variety of multicellular organisms**. The diversity of these associations, combined with their agricultural and clinical importance have made them a prominent focus of research. Of microbial genomes completed or under way, more than two-thirds are organisms that are either pathogens of humans or dependent on a close interaction with a eukaryotic host. But current databases and scientific literature present a distorted view of bacterial diversity. Estimates of bacterial diversity from various environmental sources, including the biota from animal surfaces and digestive tracts, show that **pathogens represent a very small portion of the microbial species**. Potential hosts, especially humans with their broad geographic distribution and high population densities, are constantly besieged by bacteria in the environment, but most do not cause infection.
>
> **Not only are pathogenesis and symbiosis relatively rare among bacteria species, they are derived conditions within bacteria as a whole**, as evident from the fact that bacteria existed well before their eukaryotic hosts. The appearance of the major groups of eukaryotes, whose diversification could proceed only after the origin of mitochondria by endosymbiosis, marks the initial availability of abundant suitable hosts. The mi-

tochondria themselves derive from a single lineage within the alpha subdivision of the Proteobacteria; that is to say, they are nested near the tips of the overall bacterial phylogeny. Thus, the distribution of pathogens and symbionts in numerous divergent clades of bacteria reflects the repeated and independent acquisition of this life-style.

In the second paragraph, these writers elaborate on the topic introduced in the first—that pathogenesis and symbioses are rare among bacteria—by adding that these associations evolved after the rise of eukaryotes and were derived repeatedly and independently through time.

Transition Words Revisited

Transition words (discussed in Chapter 6) can help readers move easily from one sentence or one paragraph to another. These words and phrases foretell specific arrangement patterns such as chronological order or compare and contrast. If you use the appropriate transition words in the right places, usually at the beginnings of sentences and paragraphs, your reader will be better prepared for what's to come. Here are some transition words and the arrangement patterns they foretell.

Transition words that foretell chronological order: first, second, third, initially, then, finally, in conclusion, thus, to conclude, to summarize, another, after, afterward, at last, before, presently, during, earlier, immediately, later, meanwhile, now, recently, simultaneously, subsequently.

Transition words that foretell general to specific: for example, for instance, namely, specifically, to illustrate, accordingly, consequently, hence, so, therefore, thus.

Transition words that foretell least important to most important: clearly, most importantly, the most serious, the most weighty, the foremost.

Transition words that foretell a problem and a solution: but, however, in spite of, nevertheless, nonetheless, notwithstanding, instead, still, yet.

Transition words that foretell a compare and contrast: similarly, also, in the same way, just as, so too, likewise, in comparison, however, in contrast, on the contrary, unlike, however, on the one hand . . . on the other hand.

What arrangement pattern have the following writers used to order their paragraphs? What words foretell the pattern? Check your answers with the Exercise Key in Appendix 2.

1. Previous studies have shown that animals produce different anti-predator vocalizations for aerial and terrestrial predators. Most of these studies, however, have presented these two types of predators in different ways, potentially confounding the interpretation that prey distinguish between types of predators and not their location or behavior. Our results show that chickadees do not vocally discriminate between raptors and mammals when they are presented in similar ways, and thus the "chick-a-dee" call does not refer specifically to the type of predator.

   Instead, these vocal signals likely contain information about the degree of threat that a predator represents. Maneuverability (e.g., as measured by turning radius, or radial acceleration) is extremely important in determining the outcome of predator-prey interactions and is inversely related to wingspan and body size in birds. Body size may be a good predictor of risk for chickadees: Small raptors tend to be much more maneuverable than larger raptors and likely pose a greater threat.

2. We tend to take for granted the ability of our immune systems to free our bodies of infection and prevent its recurrence. In some people, however, parts of the immune system fail. In the most severe of these immunodeficiency diseases, adaptive immunity is completely absent, and death occurs in infancy from overwhelming infection unless heroic measures are taken. Other less catastrophic failures lead to recurrent infections with particular types of pathogen, depending on the particular deficiency. Much has been learned about the functions of the different components of the human im-

mune system through the study of these immunodeficiencies, many of which are caused by inherited genetic defects.

More than 25 years ago, a devastating form of immunodeficiency appeared, the acquired immune deficiency syndrome, or AIDS, which is caused by an infectious agent, the human immunodeficiency viruses HIV-1 and HIV-2. This disease destroys T cells, dendritic cells, and macrophages bearing CD4, leading to infections caused by intracellular bacteria and other pathogens normally controlled by such cells. These infections are the major cause of death from this increasingly prevalent immunodeficiency disease.

Basic Writing Concepts

To write clearly, you don't need to know a lot of complicated grammatical terms. What you need to know are some basic writing concepts: parts of speech and elementary sentence structure. These concepts are described with examples below.[1]

Parts of Speech

Parts of speech describe the roles words play in sentences.

1. **Nouns** name persons, places, things, or ideas, and they act as subjects and objects in sentences. I focused on two kinds of nouns in this book: **concrete nouns** that refer to tangible objects such as *vertebrates*, *genes*, and DNA; and **abstract nouns** that refer to intangible ideas, emotions, or qualities such as *understanding*, *interpretation*, and *prediction*. Nouns are often introduced by articles: *a*, *an*, or *the*.

Scientists have studied *the* **arms** of spiral **galaxies.**

The nouns in this example are all concrete: *scientists*, *arms*, and *galaxies*. *Arms* is introduced by the article *the*.

2. **Pronouns** substitute for nouns. Some common pronouns are: *I*, *you*, *we*, *it*, *they*, *this*, *that*, *those*, *who*, *which*, and *what*.

They have studied **them** to determine the origin of spiral structure.

The pronoun *they* substitutes for the noun *scientists* in the previous example, while the pronoun *them* substitutes for the noun *arms*.

3. **Adjectives** describe or modify nouns or pronouns.

Many scientists have studied the **beautiful** arms of **spiral** galaxies.

The adjective *many* modifies *scientists*, *beautiful* modifies *arms*, and *spiral* modifies *galaxies*.

4. **Verbs** show action or state of being.

Many scientists have studied the beautiful arms of spiral galaxies.

The verb *have studied* describes the action.

5. **Verbals** look like verbs but function as adjectives, adverbs, or nouns.

The beautiful arms of spiral galaxies, outlined by luminous young stars, have been the focus of many studies.

In this sentence, *have been* is the verb. *Outlined* is a verbal that comes from the verb *outline*, but it functions as an adjective modifying the noun *arms*.

6. **Adverbs** modify verbs, adjectives, other adverbs, or whole sentences. They limit or define those parts of speech by answering the questions *when? how? where? how often?* and *to what extent?*

We can identify the link between density wave acceleration of molecular clouds and the subsequent birth of stars most easily and unambiguously in galaxies external to our own.

The adverbs *easily* and *unambiguously* modify the verb *can identify*. Another adverb, *most*, modifies the adverb *easily*.

7. Most **prepositions** show position in space and time. Examples are: *in, under, above, below, at, after, before, until, for, with, by,* and *on*. However, a few common prepositions do not, such as *of, as,* and *like*.

Molecular clouds spend a large fraction of their time within the arms of spiral galaxies.

8. **Conjunctions** connect parts of sentences or entire sentences. Some common conjunctions are: *and, but, or, nor, for, yet, so, however, consequently, therefore,* and *because*.

Many scientists have studied the beautiful arms of spiral galaxies outlined by luminous young stars, **but** the origin of spiral structure **and** the trigger for star formation remains unclear.

Sentence Structure

Sentence structure refers to the way sentences are formed from words, phrases, and clauses.

1. The main action word in a sentence is the **verb**, and a good way to start deciphering sentence structure is by finding it.

Many shorebirds <u>deplete</u> the prey.

The verb is *deplete*, and it describes the action.

2. The **subject** is what you write your sentence *about*. Asking "who?" or "what?" in front of the verb, usually gives you the subject.

Many <u>shorebirds</u> <u>deplete</u> the prey.

The verb is *deplete*. Who or what depletes? The answer is *shorebirds*, the subject.

3. A sentence may or may not have an **object**—a noun that *receives* the action of the verb. In most cases, asking "who?" or "what?" after the verb gives you the object (circled below).

Many <u>shorebirds</u> <u>deplete</u> the (prey).

Many shorebirds deplete what? The answer is *prey*, the object.
In the following example, the sentence has no object.

The <u>shorebirds</u> <u>flew</u>.

4. A **clause** is a group of words with a subject and a verb. There are two kinds of clauses: independent and dependent. Two or more clauses often make up a single sentence.

An **independent clause** has both subject and verb and can stand alone because it makes a complete statement. The simplest sentence is an independent clause.

When shorebirds deplete the prey in a small area, the flock must move to a fresh site.

The independent clause is the part of the sentence that can stand alone, *the flock must move to a fresh site.* It contains the verb *must move* and the subject *flock* and makes a complete statement.

A **dependent clause** has both subject and verb but cannot stand alone; it *depends* on an independent clause to make a complete statement.

When shorebirds deplete the prey in a small area, the flock must move to a fresh site.

The dependent clause is *When shorebirds deplete the prey in a small area.* Because the clause begins with the dependent word *when*, it does not make a complete statement and cannot stand alone even though it contains the verb *deplete* and the subject *shorebirds.*

Here is a list of the common words that begin dependent clauses:

after	since	whereas
although	so that	wherever
as	than	whether
because	though	whichever
as if	that	which
before	unless	while
even if	until	who
even though	what	whom
ever since	whatever	whose
how	when	why
if	whenever	where

5. **Phrases** are groups of related words without subjects and verbs. They can function in sentences as subjects, modifiers, or objects. I focused on two kinds of phrases in this book.

A **verbal phrase** consists of a verbal and the words associated with it, shown in brackets in the following sentence.

Many shorebirds (feeding in a small area) deplete the prey.

The verbal phrase, *feeding in a small area*, functions as an adjective describing the shorebirds. *Feeding* is the verbal followed by the prepositional phrase *in a small area*.

A **prepositional phrase** consists of a preposition, an object, and any modifiers, shown in parentheses in the sentence below.

Many <u>shorebirds</u> (in a small area) <u>deplete</u> the prey.

The prepositional phrase, *in a small area*, describes the shorebirds' position. *In* is the preposition, *area* is the object of the preposition, and *small* modifies *area*.

REFERENCE

1. Wilson, P. & Glazier, T. F. *The Least You Should Know about English* 8th edn (Wadsworth, 2003).

Exercise Key

These revisions are not the only answers to the exercises. You may think of others that may be better. Use these as your guide, and try the exercises several times. You will improve with practice.

EXERCISE 1

1. Processes undertaken by diverse plants and animals are responsible for such ecological actions as nutrient cycling, carbon storage, and atmospheric regulation.

The subject *Processes* is abstract.

REVISION

Diverse plants and animals cycle nutrients, store carbon, and regulate the atmosphere.

2. Declines in birth rates have been observed in many developed countries, and demographers expect that the transition to a stable population will eventually occur in many undeveloped nations as well.

The subject *Declines* is abstract.

REVISION

Demographers have observed declines in birth rates in many developed countries and expect that eventually such declines will also lead to stable populations in many undeveloped nations.

3. Variations in magmatism during rifting have been attributed to variations in mantle temperature, rifting velocity or duration, active upwelling, or small-scale convection.

The subject *Variations* is abstract.

REVISION

Magmatism <u>varies</u> during rifting for several reasons: changes in mantle temperature, rifting velocity or duration, active upwelling, or small-scale convection.

4. The <u>inability</u> of lateral variations in mantle temperature and composition, alone, to account for our observations <u><u>leads</u></u> us to propose that another influence was melt focusing.

The subject *inability* is abstract.

REVISION

<u>We</u> <u><u>could</u></u> not <u><u>account</u></u> for our observations with lateral variations in mantle temperature and composition alone. Another <u>influence</u> <u><u>was</u></u> melt focusing.

5. The <u>ability</u> of mudrock seals to prevent CO_2 leakage <u><u>is</u></u> a major concern for geological storage of anthropogenic CO_2.

The subject *ability* is abstract.

REVISION

<u>Geologists</u> <u><u>are</u></u> concerned that mudrock seals may allow anthropogenic CO_2 to leak from geological storage.

EXERCISE 2

1. Photographs from space taken by satellites <u><u>are</u></u> indicators of urbanization and just one of the <u>demonstrations</u> of the human footprint.

REVISION

Satellite photographs <u><u>indicate</u></u> the spread of urban areas and <u><u>demonstrate</u></u> the human footprint.

2. Weather variables (precipitation, temperature, and wind speed) are key factors in limiting summer habitat availability.

REVISION

Precipitation, temperature, and wind speed limit available summer habitat.

3. A risk management ranking system is the central mechanism for which prioritization of terrestrial invasive species is based.

REVISION

We rank terrestrial invasive species according to the threat they pose to the environment.

4. It is clear that Prairie Chickens are closely associated with sagebrush habitat throughout the year.

REVISION

Prairie Chickens occupy sagebrush habitat throughout the year.

5. The occurrence of freezing and thawing is an important control on cohesive bank erosion in the region.

REVISION

Freezing and thawing control cohesive bank erosion in the region.

EXERCISE 3

1. Environmentally sensitive solutions to the problems associated with continued population growth and development will require an environmentally literate citizenry.

REVISION

To develop sustainable solutions to the problems of human growth and development, we will need environmentally literate citizens.

2. Partnerships between professional teachers, scientists, non-professional science educators, and administrators are needed to improve the content and effectiveness of science education, particularly in rural areas.

REVISION

If we build partnerships between professional teachers, scientists, non-professional science educators, and administrators, we can improve the content and effectiveness of science education, particularly in rural areas.

3. Our ability to predict the spatial spread of exotic species and their transformation of natural communities is still developing.

REVISION

We still cannot predict with certainty how an exotic species will spread or transform a natural community.

4. The amount of magmatism that accompanies the extension and rupture of the continental lithosphere varies dramatically at rifts and margins around the world.

REVISION

When the continental lithosphere extends and ruptures at rifts and margins, the amount of accompanying magmatism varies dramatically.

5. The migration of melts vertically to the top of the melting region and then laterally along the base of the extended continental lithosphere would focus melts toward the eastern part of the basin.

REVISION

Melts migrate vertically to the top of the melting region, then laterally along the base of the extended continental lithosphere toward the eastern part of the basin.

6. Pre-treatment of tenocytes with different concentrations of wortmannin (1, 10, and 20 nM) for 1 h, treated with curcumin (5 μM) for 4 h, and then treated with IL-1β for 1 h, <u>inhibited</u> the IL-1β-induced NF-κB activation.

The IL-1β-induced NF-κB activation <u>was inhibited</u> by pre-treating tenocytes with wortmannin (1, 10, and 20 nM) for 1 h, followed by curcumin (5μM) for 4 h, and then IL-1β for 1h.

EXERCISE 4

1. Four big brown <u>bats</u> <u>served</u> as subjects in these experiments, two males and two females.

The subject *bats* did the serving, therefore the verb *served* is active.

2. The animals <u>were collected</u> from private homes in Maryland and <u>were housed</u> in the University of Maryland bat vivarium.

The subject *animals* received the collecting and the housing, therefore the verbs *were collected* and *were housed* are passive.

3. Bats <u>were maintained</u> at 80% of their *ad lib* feeding weight and <u>were</u> normally <u>fed</u> mealworms only during experiments.

The subject *Bats* received the maintaining and the feeding, therefore the verbs *were maintained* and *were fed* are passive.

4. We <u>exposed</u> the bats to a reversed 12h dark:12h light cycle, and we gave them free access to water.

The subject *We* did the exposing, therefore the verb *exposed* is active.

EXERCISE 5

1. For effective storage of industrial CO_2, retention times of $\sim 10^4$ yr or greater are required. (15 words)

Effective storage of industrial CO_2 requires retention times of ~10^4 yr or greater. (13 words)

2. It is hypothesized that groundwater pH must have been, on average, highest shortly before the Late Ordovician to Silurian proliferation of root-forming land plants. (24 words)

REVISION

We hypothesized that, on average, groundwater pH must have been highest shortly before the Late Ordovician to Silurian proliferation of root-forming land plants. (23 words)

3. We were compelled to rely on the SOC90 data as no further information on the occupational situation (employed vs. self-employed) or on the size of the firm was available in retrospective form. (32 words)

REVISION

We relied on the SOC90 data because we couldn't find past information on the number of people who were employed versus self-employed or on the size of the firm. (29 words)

4. Moreover, it has been demonstrated that mineral-water reactions increase the pH of groundwater even in the presence of abundant acid-producing lichens (Schatz, 1963). (24 words)

REVISION

Schatz (1963) demonstrated that mineral-water reactions increase the pH of groundwater, even in the presence of abundant acid-producing lichens. (19 words)

EXERCISE 6

1. For example, <u>expansion</u> of the extent of the winter range by <u>continued</u> <u>pioneering</u> of segments of the northern Yellowstone

elk herd northward from the park boundary and extensive use of these more northerly areas by greater numbers of elk have been coincident with acquisition and conversion of rangelands from livestock production to elk winter range.

Long words

expansion: three syllables, from Latin; *continued*: three syllables, Old French from Latin; *pioneering*: four syllables, from Old French; *boundary*: three syllables, Old French from Latin; *extensive*: three syllables, from French or Latin; *coincident*: four syllables, from Latin; *acquisition*: four syllables, from Latin; *conversion*: three syllables, Old French from Latin; *production*: three syllables, Old French from Latin

REVISION

For example, as ranchland north of the park boundary was bought and put into winter range, elk from the northern Yellowstone herd shifted north to fill it.

Short words

bought: one syllable, from Old English; *put*: one syllable, from Old English; *shift*: one syllable, from Old English; *fill*: one syllable, from Old English

2. We conclude that snag retention at multiple spatial and temporal scales in recent burns, which will be salvage-logged, is a prescription that must be implemented to meet the principles of sustainable forest management and the maintenance of biodiversity in the boreal forest.

Long words

retention: three syllables, from Old French or Latin; *multiple*: three syllables, from Latin; *spatial*: two syllables, from Latin; *temporal*: three syllables, from Old French or Latin; *prescription*: three syllables, Old French from Latin; *implemented*: four syllables, from Latin; *principles*:

three syllables, from Latin; *sustainable*: four syllables, from Latin; *management*: three syllables, from Latin; *maintenance*: three syllables, from Old French

REVISION

We found that leaving snags in salvage-logged burns helped keep biodiversity high.

Short words
find: one syllable, from Old English; *leave*: one syllable, from Old English; *help*: one syllable, from Old English; *keep*: one syllable, from Old English; *high*: one syllable, from Old English

EXERCISE 7

1. One way to assess the perceived risk of feeding in different locations is to measure the proportion of the available food a forager removes before switching to an alternative patch. All else being equal, foragers should be willing to forage longer and remove more food from a safe area than a risky one.

REVISION

We can assess the perceived risk of feeding in different patches by measuring the proportion of the available food a forager removes before switching to an alternative patch. All else being equal, foragers should be willing to forage longer and remove more food from a safe patch than a risky one.

2. Stress coping styles have been characterized as a proactive/reactive dichotomy in laboratory and domesticated animals. In this study, we examined the prevalence of proactive/reactive stress coping styles in wild-caught short-tailed mice (*Scotinomys teguina*). We compared stress responses to spontaneous singing, a social and reproductive behavior that characterizes this species.

Studies show that many animals manage stress either proactively or reactively. Here we examine whether individual wild-caught short-tailed mice (*Scotinomys teguina*) are proactive or reactive in the way they manage stress in response to spontaneous singing—a characteristic social and reproductive behavior of the species.

3. Antimicrobial resistance genes allow a microorganism to expand its ecological niche, allowing its proliferation in the presence of certain noxious compounds. From this standpoint, it is not surprising that antibiotic resistance genes are associated with highly mobile genetic elements, because the benefit to a microorganism derived from antibiotic resistance is transient, owing to the temporal and spatial heterogeneity of antibiotic-bearing environments.

Microbes can live and even proliferate in noxious environments thanks to antibiotic resistance genes. Antibiotic resistance genes are associated with highly mobile genetic elements that help microbes deal with constantly changing antibiotic-containing environments.

4. Studies of long-term outcomes in offspring exposed to maternal undernutrition and stress caused by the Dutch Hunger Winter of 1944 to 1945 revealed an increased prevalence of metabolic disease, such as glucose intolerance, obesity, and cardiovascular disease, as well as emotional and psychiatric disorders. Animal models have been developed to assess the long-term consequences of a variety of maternal challenges including under- and over-nutrition, hyperglycemia, chronic stress, and inflammation. Exposures to a wide range of insults during gestation are associated with convergent effects on fetal growth, neurodevelopment, and metabolism.

Studies show that when mothers <u>starved</u> during pregnancy in the Dutch Hunger Winter of 1944 to 1945, their children often developed metabolic diseases later in life, which included glucose intolerance, obesity, and cardiovascular disease as well as emotional and psychiatric disorders. Research on animals shows similar results: <u>starvation</u>, overeating, chronic stress, and inflammation during pregnancy influence fetal growth, neurodevelopment, and metabolism.

EXERCISE 8

1. Developing <u>regular exercise programs</u> and <u>diet regimes</u> contributes to <u>disease risk prevention</u> and <u>optimal health promotion.</u>

REVISION

Regular exercise and attention to diet help prevent disease and promote health.

2. Research focused on <u>care time deficits</u> and <u>time squeezes</u> for families has identified the persistence of <u>gendered care time burdens</u> and the sense of <u>time pressure</u> many <u>dual-earner families</u> experience around care.

REVISION

Research on families where both parents work shows that the demands of child care are stressful and still met largely by mothers.

3. There will be <u>major conservation implications</u> if <u>mercury ingestion</u> in ospreys causes <u>negative population level effects</u> either through <u>direct mortality</u> or <u>negative fecundity.</u>

If ospreys decline because ingesting mercury either kills them directly or lowers their breeding success, we face a serious conservation problem.

EXERCISE 9

1. One of the well-researched immunoregulatory functions of probiotics is the induction of cytokine production. In particular, the induction of IL-10 and IL-12 production by probiotics has been studied intensively, because the balance of IL-10/IL-12 secreted by macrophages and dendritic cells in response to microbes is crucial for determination of the direction of the immune response. IL-10 is an anti-inflammatory cytokine and is expected to improve chronic inflammation, such as that of inflammatory bowel disease and autoimmune disease. IL-10 downregulates phagocytic and T cell functions, including the production of pro-inflammatory cytokines, such as IL-12, TNF-α, and IFN-γ, that control inflammatory responses. IL-10 promotes the development of regulatory T cells for the control of excessive immune responses. In contrast, IL-12 is an important mediator of cell-mediated immunity and is expected to augment the natural immune defense against infections and cancers. IL-12 stimulates T cells to secrete IFN-γ, promotes Th1 cell development, and, directly or indirectly, augments the cytotoxic activity of NK cells and macrophages. IL-12 also suppresses redundant TH2 cell responses for the control of allergy.

The writers have introduced two technical terms their readers may not be familiar with: IL-10 (Interleukin-10) and IL-12 (Interleukin-12). They have made these terms understandable to an audience familiar with undergraduate-level immunology. They tell us that both are cytokines and describe the roles they play in the immune system. IL-10 is an anti-inflammatory. It downregulates phagocytic and T cell functions, and promotes the development of regulatory T cells. IL-12 is involved in cell-mediated immunity, stimulates T cells, and suppresses redundant Th2 cell responses.

EXERCISE 10

1. While a growing body of research indicates that large herbivores as a group can exert strong indirect effects on co-occurring species, there are comparatively few examples of strong community-wide impacts from individual large herbivore species. (37 words)

While a growing body of is unnecessary detail; *indicates* is a hedge; *as a group* is redundant; *can exert strong indirect effects* can be said in fewer words; *comparatively few* is a hedge; *individual large herbivore species* is a string of two nouns and two adjectives.

REVISION

Research shows that large herbivores can indirectly influence co-occurring species, but few studies focus on a single species of large herbivore and how it affects the whole community. (28 words)

2. Small mammal species diversity increased in exclosures relative to controls, while survivorship showed no significant trends. (16 words)

Small mammal species diversity is a string of one adjective and three nouns; *species* is redundant; *no significant trends* is negative.

REVISION

Diversity of small mammals increased in exclosures relative to controls, while survivorship stayed the same. (15 words)

3. In this essay, I will be looking at how higher summer temperatures cause quicker soil and plant evaporation. We all know that climate change has caused elevated temperatures in the Northwest throughout the spring and summer months. We also know that these record-breaking temperatures have the effect of

quickly and easily desiccating soil and drying out plant foliage so that it is more flammable. Understandably then, when lightning strikes this very combustible environment, a spark can very quickly turn into a widespread blaze. (83 words)

~~In this essay I will be looking at how,~~ We all know that, We also know that these, and Understandably then are unnecessary metadiscourse; higher summer temperatures and record-breaking temperatures is repetitive and could be said in fewer words; more flammable and very combustible environment is repetitive; widespread blaze could be said in fewer words.

REVISION

Due to climate change, spring and summer temperatures in the Northwest are becoming warmer. Warmer temperatures dry out both soils and plant foliage, which are then more prone to wildfires. (30 words)

4. Zimbabwean undocumented migrants are shown to be marginalized and vulnerable with limited transnational citizenship. (14 words)

Zimbabwean undocumented migrants is an unwieldy string of two long adjectives and one noun; are shown to be is a passive verb and suggest an unknown observer—a type of metadiscourse.

REVISION

Undocumented migrants from Zimbabwe are marginalized and vulnerable with limited transnational citizenship. (12 words)

5. When the lithosphere extends and rifts along continental margins, magma is produced in varying quantities. Widely spaced geophysical transects show that rifting along some continental margins can transition from magma-poor to magma-rich. Our wide-angle seismic data from the Black Sea provide the first direct observations of such a transition. This transition coincides with a transform fault and is abrupt, occurring over only

~20–30 km. This abrupt transition cannot be explained solely by gradual along-margin variations in mantle properties, since these would be expected to result in a smooth transition from magma-poor to magma-rich rifting over hundreds of kilometers. We suggest that the abruptness of the transition results from the 3-D migration of magma into areas of greater extension during rifting, a phenomenon that has been observed in active rift environments such as mid-ocean ridges. (133 words)

This paragraph is good as is.

6. The empirical data presented in this article reveal a segmented labor market and exploitation, with undocumented migrants not benefiting from international protection, human rights, nation state citizenship rights, or rights associated with the more recent concepts of post-national and transnational citizenship. (41 words)

The empirical data presented in this article is unnecessary metadiscourse; *not benefiting* is negative.

REVISION

The segmented labor market we describe exploits undocumented migrants. These people lack international protection, human rights, nation-state citizenship, and rights associated with the more recent concepts of postnational and transnational citizenship. (30 words)

7. The systemic immune response in *Drosophila* is mediated by a battery of antimicrobial peptides produced largely by the fat body, an insect organ analogous to the mammalian liver. These peptides lyse microorganisms by forming pores in their cell walls. Functionally, the antimicrobial peptides fall into three classes depending on the pathogen specificity of their lytic activity. Thus, Drosomycin is a major antifungal peptide, whereas Diptericin is active against gram-negative bacteria, and Defen-

sin works against gram-positive bacteria. Interestingly, infection of *Drosophila* with different classes of pathogens leads to preferential induction of the appropriate group of antimicrobial peptides.

The transition words are: *functionally, thus, whereas,* and *interestingly.* These words help the reader follow the writer's train of thought, explaining the role of antimicrobial peptides, clarifying the items in the list, and pointing out interesting facts.

8. Introgressive hybridization is most commonly observed in zones of geographical contact between otherwise allopatric taxa. Studies of such zones have provided important insights into the evolutionary process and have helped resolve part of the debate about fitness of hybrids. In many cases, most hybrid genotypes tend to be less fit than are the parental genotypes in parental habitats, owing either to endogenous or exogenous selection or both. However, theory predicts that some can be of equal or superior fitness in new habitats and, occasionally, even in parental habitats.

Many principles of plain English are evident in the example: subjects are followed closely by verbs; verbs are active except for one passive—*is most commonly observed*, which allows the author to use *Introgressive hybridization* as the subject; needless metadiscourse is absent, negatives are absent, repetition and excessive detail are absent. The writing is concise.

EXERCISE 11

1. Unfortunately, as noted 40 years ago, few students experience the thrill of doing field science because they are rarely allowed to leave the confines of the classroom to become immersed in field-based science.

Here the writer places old information at the end of the sentence, *to become immersed in field-based science.*

Unfortunately, as noted 40 years ago, few students experience the thrill of doing field science because they are rarely allowed to leave the confines of the classroom.

2. Bank erosion rates along the South River in Virginia increased by factors of 2–3 after 1957. Increased bank erosion rates cannot be explained by changes in the intensity of either freeze-thaw or storm intensity, and changes in the density of riparian trees should have decreased erosion rates.

The subjects in these two sentences are consistent, *bank erosion rates*; however, old information is placed incorrectly at the end of the second sentence as *decreased erosion rates*. The second sentence also contains a negative, *cannot be explained*.

After 1957, bank erosion in Southern Virginia increased by 2–3 times. These increases have little to do with the severity of freeze-thaw cycles, or with the intensity of storms, and changes in the density of riparian trees should have had a stabilizing effect.

3. Students majoring in science often believe they can escape the intensive writing and presentations that their peers in the humanities and social sciences must do. However, science is a collective human endeavor whose success hinges upon effective communication, both written and oral. Even if findings are ground breaking, they are potentially worthless if they can't be shared with others in a clear and engaging way. Teaching undergraduate science students to effectively communicate is therefore an essential goal.

The subjects in this paragraph, if we include their modifiers, are fairly consistent, *students (majoring in science)*, *science*, *findings* (findings in science is understood), and *teaching (science students)*. The first three sentences end with unnecessary words that hide the new information.

Science students often believe they can escape the intensive writing and presentations of their peers in the humanities and social sciences. However, science is a collective human endeavor whose success hinges upon effective communication. Even if findings are ground breaking, they are potentially worthless if they can't be shared. Therefore, teaching undergraduate science students to communicate effectively is an essential goal.

4. Climate plays an important part in determining the average numbers of a species, and periodical seasons of extreme cold or drought, I believe to be the most effective of all checks. I estimated that the winter of 1854–55 destroyed four-fifths of the birds in my own grounds; and this is a tremendous destruction, when we remember that ten per cent is an extraordinarily severe mortality from epidemics with man. The action of climate seems at first sight to be quite independent of the struggle for existence; but in so far as climate chiefly acts in reducing food, it brings on the most severe struggle between the individuals, whether of the same or of distinct species, which subsist on the same kind of food. Even when climate, for instance extreme cold, acts directly, it will be the least vigorous, or those which have got least food through the advancing winter, which will suffer most.

These sentences were written by Charles Darwin from his famous work The Origin of Species. Each is a good example of new and old information placed correctly.

EXERCISE 12

1. Central to this deficit has been the rising average age of the nursing workforce and the decline in the number of hours worked; fewer nurses are working standard full-time hours (35–44 hours per week) and 44 percent work part-time.

There are two lists in this sentence, and neither is parallel. The items in the first list are:

the rising average age of the nursing workforce, and
the decline in the number of hours worked.

To revise, change *nursing workforce* to *nurses*, and *decline* to *declining*. This gives you:

the rising average age of nurses and
the declining number of hours worked.

The items in the second list are:

fewer nurses are working standard full-time hours (35–44 hours
    per week), and
44 percent work part-time.

To revise, delete *standard* and *hours*, add *more nurses* and *working*, and shift *44 percent* to the end. This gives you:

fewer nurses are working full-time (35–44 hours per week), and
more nurses are working part-time (44 percent).

The second *nurses* is implied and could be omitted.

REVISION

Central to this deficit has been the rising average age of nurses and the declining number of hours worked; fewer nurses are working full-time (35–44 hours per week), and more are working part-time (44 percent).

2. The problem of finding the optimal strokes of hypothetical microswimmers has drawn a lot of attention in recent years. Problems that have been solved include the optimal stroke pattern of Purcell's three-link swimmer, an ideal elastic flagellum, a shape-changing body, a two- and a three-sphere swimmer, and a spherical squirmer.

The new information that introduces the list, *The problem of finding the optimal strokes of hypothetical microswimmers*, is at the beginning of the first sentence, and the list is not parallel:

the optimal stroke pattern of Purcell's three-link swimmer,
~~an ideal elastic flagellum,~~
a shape-changing body,
a two- and three-sphere swimmer, and
a spherical squirmer.

Put the information that introduces the list, *optimal strokes of hypothetical microswimmers*, at the end of the first sentence. Begin the second sentence with *Optimal strokes* as the subject, and introduce each item on the list with the same word, *for*. (It may not be necessary to repeat the *for* after the first item, as it is understood.) This gives you:

for Purcell's three-link swimmer,
an ideal elastic flagellum,
a shape-changing body,
a two- and three-sphere swimmer, and
a spherical squirmer.

REVISION

In recent years, much attention has focused on finding the optimal strokes of hypothetical microswimmers. Optimal strokes have been found for Purcell's three-link swimmer, an ideal elastic flagellum, a shape-changing body, a two- and three-sphere swimmer, and a spherical squirmer.

3. Cilia are hair-like protrusions that beat in an asymmetric fashion to pump the fluid in the direction of their effective stroke. They propel certain protozoa, such as *Paramecium*, and also fulfill a number of functions in mammals, including mucous clearance from airways, left-right asymmetry determination, and transport of an egg cell in fallopian tubes.

The new information in the second sentence (*propel certain protozoa,* and *fulfill a number of functions*) is at the beginning and middle of the sentence. *Functions* is abstract, and the list is not parallel:

mucous clearance in airways,
left-right asymmetry determination, and
transport of an egg cell in fallopian tubes.

Introduce the list with a short sentence, ending it with *different organisms.* Give each kind of organism their own sentence, beginning with similar words: *In protozoa . . .* and *In mammals . . .* ending with the new information. Make the items in the list parallel by introducing each with a present tense verb:

clear mucous from airways,
determine left-right asymmetry, and
transport egg cells in fallopian tubes.

REVISION

Cilia are hair-like protrusions that beat asymmetrically to pump fluid in the direction of their effective stroke. This pumping is put to an astonishing variety of uses by different organisms. In protozoa such as *Paramecium*, cilia provide the primary power for movement. In mammals, cilia clear mucous from airways, determine left-right asymmetry, and transport egg cells in fallopian tubes.

4. Integrons consist of three elements: an attachment site where the horizontally acquired sequence is integrated; a gene encoding a site-specific recombinase (that is, integrase); and a promoter that drives expression of the incorporated sequence.

The list is not parallel:

an attachment site where the horizontally acquired sequence is
    integrated;

a gene encoding a site-specific recombinase (that is, integrase);
and

a promoter that drives expression of the encorporated sequence.

Follow each item with a dependent clause starting with the same word, that, followed by a present-tense verb, which gives you:

an attachment site that locates the integration of the horizontally acquired sequence;

a gene that encodes a site-specific recombinase (that is, integrase);
and

a promoter that drives expression of the incorporated sequence.

REVISION

Integrons consist of three elements: an attachment site that locates the integration of the horizontally acquired sequence; a gene that encodes a site-specific recombinase (that is, integrase); and a promoter that drives expression of the incorporated sequence.

5. North American (NA)-EEEV strains cause periodic outbreaks of mosquito-borne encephalitis in humans and equines, are highly neurovirulent, and, in comparison with related Venezuelan equine encephalitis virus (VEEV) and western equine encephalitis virus (WEEV), cause far more severe encephalitic disease in humans.

The list is parallel. Each of the three items start with a present-tense verb: *cause, are,* and *cause*. Each verb is followed by an object and its modifier(s): *periodic outbreaks, highly neurovirulent,* and *far more severe encephalitic disease*.

EXERCISE 13

1. In order to unravel the mode of action of neuronal networks, a neurobiologist's dream would be not only to be able to monitor neuronal activity but also to have control over distinct sets

of neurons and to be able to manipulate their activity and observe the effect on behavior. (49 words) This idea is not new. (5 words) As the activity of a neuron is based on the depolarization of its cell membrane, neuronal activity can be induced by an experimenter using stimulation electrodes by which the cell membrane can be artificially depolarized or hyperpolarized. (37 words) Although stimulation electrodes have served, and continue to serve, neuroscientists well for decades, limitations of this invasive approach are obvious. (20 words)

This example shows a wide variety of sentence lengths ranging from five to 49 words. The average number of words per sentence is 28, which is long. You can reduce the sentence length by deleting needless words. For example, the first sentence:

In order to unravel the mode of action of neuronal networks, a neurobiologist's dream would be not only to be able to monitor neuronal activity but also to have control over distinct sets of neurons and to be able to manipulate their activity and observe the effect on behavior. (49 words)

could be revised to:

Neurobiologists dream of being able to monitor neuronal activity and control and manipulate distinct sets of neurons, while at the same time observing how these changes affect behavior. (28 words)

2. The extrapolation from *in vitro* measurements to the *in vivo* behavior of proteins is hampered by extremely high (300–400 mg/mL) intracellular macromolecular concentrations in the cell, i.e. crowding, which is one of the most important factors that influences the structure and function of proteins under physiological conditions. (47 words)

This sentence is too long. Break it up and place important information at the ends of the sentences.

One of the most important factors influencing the structure and function of proteins under physiological conditions is crowding. (18 words) Crowding occurs when extremely high (300–400 mg/mL) concentrations of intracellular macromolecules occur in the cell, making it hard to extrapolate from *in vitro* measurements to the proteins' *in vivo* behavior. (30 words)

3. Some of the confusion about the role of hybridization in evolutionary diversification stems from the contradiction between a perceived necessity for cessation of gene flow to enable adaptive population differentiation on the one hand, and the potential of hybridization for generating adaptive variation, functional novelty, and new species on the other. (52 words)

This sentence is too long. Break it up so that each sentence has one, or at most two, main ideas.

REVISION

Whether interspecific hybridization is important as a mechanism that generates biological diversity is controversial. (14 words) While some authors see hybrids as a source of genetic variation, functional novelty, and new species, others argue that reduced fitness would typically render hybrids an evolutionary dead end. (29 words)

EXERCISE 14

1. Males of *O. acuminatus* [dung beetles] employed two very different tactics to encounter and mate with females: they either attempted to monopolize access to a female by guarding the entrance to her tunnel (guarding), or they attempted to bypass guarding males (sneaking). Guarding behavior entailed remaining inside a tunnel with a female, and fighting intruding males over possession of the tunnel. Guarding males blocked tunnel entrances and periodically "patrolled" the length of the tunnel.

Rival males could gain possession of a tunnel only by forcibly evicting the resident male, and both fights and turnovers were frequent. Fights over tunnel occupancy entailed repeated butting, wrestling and pushing of opponents, and fights continued until one of the contestants left the tunnel.

The first sentence is the issue. In it, the writer introduces the two tactics dung beetles employ to meet and mate with females: guarding and sneaking. A description of guarding behavior follows in the development.

2. One prerequisite for the maintenance of dimorphism is that organisms experience a fitness tradeoff across environments. If animals encounter several discrete environment types, or ecological or behavioral situations, and these different environments favor different morphologies, then distinct morphological alternatives can evolve within a single population—each specialized for one of the different environments. Such fitness tradeoffs have been demonstrated for several dimorphic species. For example, soft and hard seed diets have favored two divergent bill morphologies within populations of African finches, and high and low levels of predation have favored alternative shell morphologies in barnacles, and spined and spineless morphologies in rotifers and *Daphnia*. It is possible that the alternative reproductive tactics characterized in this study produce a similar situation in *O. acuminatus*. If guarding and sneaking behaviors favor horned and hornless male morphologies, respectively, then the reproductive behavior of males may have contributed to the evolution of male horn length dimorphism in this species.

The first three sentences make up the issue. The writer introduces ideas about the evolution and maintenance of dimorphism in discrete environments. The last sentence of the issue introduces what will appear in the development: a list of examples of *several dimorphic species* whose fitness tradeoffs have been demonstrated.

3. In recent times, the origin of the adaptive immune response has been uncovered. It turns out that the two recombinase-activated genes are encoded in a short stretch of DNA, in opposite orientations and lacking exons. This suggested an origin in a retroposon, as did the presence of the recognition signal sequences that lie 3' of all V gene segments and 5' of all J gene segments. This hypothesis was tested *in vitro* and shown to be true. Other processes expand diversity tremendously, such as the generation of D gene segments in the first chain to rearrange, the nucleotide-adding enzyme TdT that inserts nucleotides in the junctions of V-D-J junctions, and somatic hypermutation.

The first sentence is the issue. The writer introduces the idea of the origin of the adaptive immune response and proceeds to give details of the genetic origin in the development.

4. Despite a consistent correlation between genome size and the obligate association with host cells, genome reduction is not simply an adaptive response to living within hosts. Instead, the trend toward large-scale gene loss reflects a lack of effective selection for maintaining genes in these specialized microbes. Because the host presents a constant environment rich in metabolic intermediates, some genes are rendered useless by adoption of a strictly symbiotic or pathogenic life-style. These superfluous sequences are eliminated through mutational bias favoring deletions, a process apparently universal in bacterial lineages. Thus, all of the fully sequenced small genomes display a pattern of loss of biosynthetic pathways, such as those for amino acids that can be obtained from the host cytoplasm.

The first two sentences are the issue. The writer sets up a premise that he then contradicts: *Despite a constant correlation . . . genome reduction is not simply an adaptive response . . .* and *Instead . . . large scale genome loss . . . reflects a lack of . . . selection for maintaining genes . . .* The development explains how this lack of selection occurs and the conclusion summarizes.

5. Unlike pathogens, symbionts may devote part of their genomes to processes that are more directly beneficial to the host rather than to the bacterial cell itself. *Buchnera* retains and even amplifies genes for the biosynthesis of amino acids required by hosts, devoting almost 10% of its genome to these pathways, which are missing from pathogens with similarly small genomes. Because of their fastidious growth requirements, the biological role of obligately associated symbionts can rarely be determined experimentally. However, genome comparisons can provide a means for determining their functions in hosts. Such future research should reveal, for example, whether the endosymbionts of blood-feeding hosts, such as *Wigglesworthia glossinia* in tsetse flies, retain pathways for biosynthesis of vitamins absent from blood, whether the symbiont *Vibrio fischeri* provides functions other than bioluminescence to its squid host, and whether the mutualistic *Wolbachia* of filarial nematodes contain genes for host benefit that are absent in the parasitic *Wolbachia* of arthropods.

The first sentence is the issue. In it, the writer introduces the idea that symbionts may devote part of their genomes to processes that benefit their hosts. The development gives an example of such a symbiont. The conclusion is speculative, describing what future research on obligate symbionts should reveal.

EXERCISE 15

1. Previous studies have shown that animals produce different antipredator vocalizations for aerial and terrestrial predators. Most of these studies, however, have presented these two types of predators in different ways, potentially confounding the interpretation that prey distinguish between types of predators and not their location or behavior. Our results show that chickadees do not vocally discriminate between raptors and mammals when they are presented in similar ways, and thus the "chick-a-dee" call does not refer specifically to the type of predator.

Instead, these vocal signals likely contain information about the degree of threat that a predator represents. Maneuverability (e.g., as measured by turning radius, or radial acceleration) is extremely important in determining the outcome of predator-prey interactions and is inversely related to wingspan and body size in birds. Body size may be a good predictor of risk for chickadees: Small raptors tend to be much more maneuverable than larger raptors and likely pose a greater threat.

In the first paragraph, the writers describe a problem with previous studies. The way predators were presented to prey animals (as either aerial or terrestrial) in vocalization experiments may confound the results. In the second paragraph, the writers propose a solution and show that if predators are presented in the same way, the prey regard the predator's body size and maneuverability as more important than whether it is aerial or terrestrial.

The writers start the second paragraph with the word instead, which signals that the paragraph will present another way to approach the problem.

2. We tend to take for granted the ability of our immune systems to free our bodies of infection and prevent its recurrence. In some people, however, parts of the immune system fail. In the most severe of these immunodeficiency diseases, adaptive immunity is completely absent, and death occurs in infancy from overwhelming infection unless heroic measures are taken. Other less catastrophic failures lead to recurrent infections with particular types of pathogen, depending on the particular deficiency. Much has been learned about the functions of the different components of the human immune system through the study of these immunodeficiencies, many of which are caused by inherited genetic defects.

More than 25 years ago, a devastating form of immunodeficiency appeared, the acquired immune deficiency syndrome, or AIDS, which is caused by an infectious agent, the human immunodeficiency viruses HIV-1 and HIV-2. This disease destroys

T cells, dendritic cells, and macrophages bearing CD4, leading to infections caused by intracellular bacteria and other pathogens normally controlled by such cells. These infections are the major cause of death from this increasingly prevalent immunodeficiency disease.

In these two paragraphs, the writers go from general information to specific. The first paragraph is a short review of immunodeficiency disease. The second paragraph focuses on one kind of immunodeficiency disease—AIDS.

Although no word foretells the pattern of this arrangement, the writers use a more subtle technique to link these two paragraphs. They end the first paragraph with an indirect reference to time: *Much has been learned about the functions of the different components of the human immune system through the study of these immunodeficiencies.* Then they begin the second paragraph with another reference to time that echoes the first: *More than 25 years ago, a devastating form of immunodeficiency appeared, the acquired immune deficiency syndrome, or AIDS.*

Index

Page numbers for definitions are in boldface.